An Introduction to the
Geometry *of*
Stochastic Flows

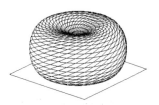

An Introduction to the
Geometry *of*
Stochastic Flows

Fabrice Baudoin
Université Paul Sabatier, France

Imperial College Press

Published by

Imperial College Press
57 Shelton Street
Covent Garden
London WC2H 9HE

Distributed by

World Scientific Publishing Co. Pte. Ltd.
5 Toh Tuck Link, Singapore 596224
USA office: 27 Warren Street, Suite 401-402, Hackensack, NJ 07601
UK office: 57 Shelton Street, Covent Garden, London WC2H 9HE

British Library Cataloguing-in-Publication Data
A catalogue record for this book is available from the British Library.

QA274.23
.B384
2004x

05766769l

AN INTRODUCTION TO THE GEOMETRY OF STOCHASTIC FLOWS

Copyright © 2004 by Imperial College Press

ISBN 1-86094-481-7

Printed in Singapore by World Scientific Printers (S) Pte Ltd

Au vent d'autan...

Preface

The aim of the present text is to provide a self-contained introduction to the local geometry of the stochastic flows associated with stochastic differential equations.

The point of view we want to develop is that the local geometry of any stochastic flow is determined very precisely and explicitly by a universal formula referred to as the Chen-Strichartz formula. The natural geometry associated with the Chen-Strichartz formula is the sub-Riemannian geometry whose main tools are introduced throughout the text. By using the connection between stochastic flows and partial differential equations, we apply this point of view to the study of hypoelliptic operators written in Hörmander's form.

Many results contained in this text stem from my stay at the Technical University of Vienna where I had the great pleasure to discuss passionately with Josef Teichmann. I learnt a lot from him and I thank him very warmly. I also would like to take this opportunity to thank Nicolas Victoir for reading early drafts of various parts of the text and for his valuable suggestions.

F. Baudoin,
Toulouse, June 2004

Contents

Chapter 1

Formal Stochastic Differential Equations

The goal of this first chapter is to establish the Chen-Strichartz formula which, in a way, is a cornerstone of this book. This formula is universal and determines very precisely and explicitly the local structure of any stochastic flow. To derive this formula, it is quite convenient to work in an abstract and formal setting, in which we do not have to care about convergence questions.

The reader which is not so familiar with the theory of stochastic differential equations and vector fields is invited read the Appendices A and B which are included at the end of the book.

1.1   Motivation

Let us consider a stochastic differential equation on $\mathbb{R}^n$ of the type

$$X_t^{x_0} = x_0 + \sum_{i=1}^{d} \int_0^t V_i(X_s^{x_0}) \circ dB_s^i, \ t \geq 0, \tag{1.1}$$

where:

(1) $x_0 \in \mathbb{R}^n$;
(2) $V_1, ..., V_d$ are C^∞ bounded vector fields on $\mathbb{R}^n$;
(3) $\circ$ denotes Stratonovitch integration;
(4) $(B_t)_{t \geq 0} = (B_t^1, ..., B_t^d)_{t \geq 0}$ is a d-dimensional standard Brownian motion.

Let $f : \mathbb{R}^n \to \mathbb{R}$ be a smooth function and denote by $(X_t^{x_0})_{t \geq 0}$ the solution of (2.1) with initial condition $x_0 \in \mathbb{R}^n$. First, by Itô's formula, we have

$$f(X_t^{x_0}) = f(x_0) + \sum_{i=1}^d \int_0^t (V_i f)(X_s^{x_0}) \circ dB_s^i, \ t \geq 0.$$

Now, a new application of Itô's formula to $V_i f(X_s^x)$ leads to

$$f(X_t^{x_0}) = f(x_0) + \sum_{i=1}^d (V_i f)(x_0) B_t^i + \sum_{i,j=1}^d \int_0^t \int_0^s (V_j V_i f)(X_u^{x_0}) \circ dB_u^j \circ dB_s^i.$$

We can continue this procedure to get after N steps

$$f(X_t^{x_0}) = f(x_0) + \sum_{k=1}^N \sum_{I=(i_1,\ldots i_k)} (V_{i_1} \ldots V_{i_k} f)(x_0) \int_{\Delta^k[0,t]} \circ dB^I + \mathbf{R}_N(t),$$

for some remainder term $\mathbf{R}_N$, where we used the notations:

(1)

$$\Delta^k[0,t] = \{(t_1, \ldots, t_k) \in [0,t]^k, t_1 \leq \ldots \leq t_k\};$$

(2) If $I = (i_1, \ldots i_k) \in \{1, \ldots, d\}^k$ is a word with length k,

$$\int_{\Delta^k[0,t]} \circ dB^I = \int_{0 \leq t_1 \leq \ldots \leq t_k \leq t} \circ dB_{t_1}^{i_1} \circ \ldots \circ dB_{t_k}^{i_k}.$$

If we dangerously do not care about convergence questions (these questions are widely discussed in [Ben Arous (1989b)]), it is tempting to let $N \to +\infty$ and to assume that $\mathbf{R}_N \to 0$. We are thus led to the nice (but formal!) formula

$$f(X_t^{x_0}) = f(x_0) + \sum_{k=1}^{+\infty} \sum_{I=(i_1,\ldots i_k)} (V_{i_1} \ldots V_{i_k} f)(x_0) \int_{\Delta^k[0,t]} \circ dB^I. \qquad (1.2)$$

We can rewrite this formula in a more convenient way. Let Φ_t be the stochastic flow associated with the stochastic differential equation (2.1). There is a natural action of Φ_t on smooth functions: The pull-back action given by

$$(\Phi_t^* f)(x_0) = (f \circ \Phi_t)(x_0) = f(X_t^{x_0}).$$

The formula (1.2) shows then that we have the following formal development for this action

$$\Phi_t^* = \mathbf{Id} + \sum_{k=1}^{+\infty} \sum_{I=(i_1,\dots i_k)} V_{i_1}\dots V_{i_k} \int_{\Delta^k[0,t]} \circ dB^I. \qquad (1.3)$$

Though this formula does not make sense from an analytical point of view, at least, it shows that the *probabilistic information* contained in the stochastic flow associated with the stochastic differential equation (1.1) is given by the set of Stratonovitch chaos $\int_{\Delta^k[0,t]} \circ dB^I$. What is a priori less clear is that the *algebraic information* which is relevant for the study of Φ_t^* is given by the structure of the Lie algebra generated by the $V_i's$, and this is precisely this aspect we want to stress in this chapter which is devoted to the study of formal objects like

$$\mathbf{Id} + \sum_{k=1}^{+\infty} \sum_{I=(i_1,\dots i_k)} V_{i_1}\dots V_{i_k} \int_{\Delta^k[0,t]} \circ dB^I.$$

Such objects and their relations with flows seem to appear the first time in the works of K.T. Chen [Chen (1957)], [Chen (1961)].

1.2 The signature of a Brownian motion

Let us denote by $\mathbb{R}[[X_1, ..., X_d]]$ the **non-commutative** algebra of formal series with d indeterminates.

Definition 1.1   The signature of a d-dimensional standard Brownian motion $(B_t)_{t\geq 0}$ is the element of $\mathbb{R}[[X_1, ..., X_d]]$ defined by

$$S(B)_t = 1 + \sum_{k=1}^{+\infty} \sum_{I=(i_1,\dots i_k)} X_{i_1}\dots X_{i_k} \int_{\Delta^k[0,t]} \circ dB^I, \ t \geq 0.$$

Remark 1.1   *We define the signature by using Stratonovitch's integrals because we keep in mind the connection with stochastic flows which appeared with formula (1.3). Nevertheless, it is possible to define a signature by using Itô's integrals. The link between these two signatures is given in Proposition 1.2 below.*

Remark 1.2   *In the same way, it is of course also possible to define the signature of a general semimartingale.*

Observe that the signature hence defined is the solution of the *formal* stochastic differential equation

$$S(B)_t = 1 + \sum_{i=1}^{d} \int_0^t S(B)_s X_i \circ dB_s^i, \ t \geq 0. \tag{1.4}$$

Such linear equations appear in the study of Brownian motions on Lie groups. Indeed, let $\mathbb{G}$ be a Lie group with Lie algebra $\mathfrak{g}$.

Definition 1.2　A process $(X_t)_{t \geq 0}$ with values in $\mathbb{G}$ is called a (left) Brownian motion on $\mathbb{G}$ if:

(1) $(X_t)_{t \geq 0}$ is continuous;
(2) for each $s \geq 0$, the process $\left(X_s^{-1} X_{t+s}\right)_{t \geq 0}$ is independent of the process $(X_u)_{0 \leq u \leq s}$;
(3) for each $s \geq 0$, the processes $\left(X_s^{-1} X_{t+s}\right)_{t \geq 0}$ and $(X_t)_{t \geq 0}$ are identical in law.

In a general way, one can construct Brownian motions on Lie groups by solving differential equations. Let us consider $V_1, ..., V_d \in \mathfrak{g}$. As explained in Appendix B, $V_1, ..., V_d \in \mathfrak{g}$ can be seen as left invariant vector fields on $\mathbb{G}$, so that we can consider the following stochastic differential equation

$$X_t = 1_{\mathbb{G}} + \sum_{i=1}^{d} \int_0^t V_i(X_s) \circ dB_s^i, \ t \geq 0, \tag{1.5}$$

where $(B_t)_{t \geq 0}$ is a standard Brownian motion on $\mathbb{R}^d$. For instance, if $\mathbb{G}$ is a linear group of matrices, equation (1.5) can be rewritten

$$X_t = 1_{\mathbb{G}} + \sum_{i=1}^{d} \int_0^t X_s V_i \circ dB_s^i.$$

It is easily seen that there is a unique solution $(X_t)_{t \geq 0}$ to the stochastic differential equation (1.5), and this solution is a (left) Brownian motion on $\mathbb{G}$. The process $(X_t)_{t \geq 0}$ is called a lift of $(B_t)_{t \geq 0}$ in $\mathbb{G}$. It is interesting to note that, conversely, each Brownian motion on $\mathbb{G}$ is solution of a stochastic differential equation

$$X_t = X_0 + \int_0^t V_0(X_s) ds + \sum_{i=1}^{d} \int_0^t V_i(X_s) \circ dB_s^i, \ t \geq 0,$$

where $V_0, V_1, ..., V_d$ are left-invariant vector fields on $\mathbb{G}$; for further details on this, we refer to [Hunt (1958)] and [Yosida (1952)].

With this in mind, we interpret now $\mathbb{R}[[X_1, ..., X_d]]$ as the universal enveloping algebra of the free Lie algebra with d generators $\mathfrak{f}_d$. So, with this interpretation, at the formal level the signature of $(B_t)_{t \geq 0}$ can be interpreted as a lift of $(B_t)_{t \geq 0}$ in the formal object $\exp(\mathfrak{f}_d)$.

On the other hand, the first section of this chapter has shown that the pull-back action on functions of the stochastic flow $(\Phi_t)_{t \geq 0}$ associated with the stochastic differential equation

$$X_t^{x_0} = x_0 + \sum_{i=1}^{d} \int_0^t V_i(X_s^{x_0}) \circ dB_s^i, \ t \geq 0,$$

solves formally the stochastic differential equation

$$\Phi_t^* = \mathbf{Id} + \sum_{i=1}^{d} \int_0^t \Phi_s^* V_i \circ dB_s^i,$$

so that $(\Phi_t^*)_{t \geq 0}$ can formally be seen as a lift of $(B_t)_{t \geq 0}$ in the formal object $\exp(\mathfrak{L}(V_1, ..., V_d))$ where $\mathfrak{L}(V_1, ..., V_d)$ is the Lie algebra generated by $V_1, ..., V_d$.

Therefore, since $\mathfrak{f}_d$ is a universal object in the theory of Lie algebras, the signature appears as a universal object in the theory of stochastic flows. In particular, if we do not care about convergence questions, any algebraic formula concerning the signature of $(B_t)_{t \geq 0}$ can be applied to study the stochastic flow associated with **any** stochastic differential equation driven by $(B_t)_{t \geq 0}$. As it will be seen in the next chapters, one of the most illuminating example in this direction is certainly the universality of the Chen-Strichartz formula; an other example is given by the expectation of the signature (see the end of the chapter), a purely algebraic object, which *explains* in a different way than the usual one, the Markov property shared by any process that solves a stochastic differential equation driven by Brownian motions.

We have a fundamental flow property for the signature which stems directly from the following key but simple relations, known as the Chen's relations since the seminal work [Chen (1957)].

Lemma 1.1 *For any word* $(i_1, ..., i_n) \in \{1, ..., d\}^n$ *and any* $0 < s < t$,

$$\int_{\Delta^n[0,t]} \circ dB^{(i_1, ..., i_n)} = \sum_{k=0}^{n} \int_{\Delta^k[0,s]} \circ dB^{(i_1, ..., i_k)} \int_{\Delta^{n-k}[s,t]} \circ dB^{(i_{k+1}, ..., i_n)},$$

where we used the following notations:

(1)

$$\int_{\Delta^k[s,t]} odB^{(i_1,\dots,i_k)} = \int_{s\le t_1\le\dots\le t_k\le t} odB^{i_1}_{t_1} \circ \dots \circ dB^{i_k}_{t_k};$$

(2) if I is a word with length 0, then $\int_{\Delta^0[0,t]} odB^I = 1$.

Proof. It follows readily by induction on n by noticing that

$$\int_{\Delta^n[0,t]} odB^{(i_1,\dots,i_n)} = \int_0^t \left(\int_{\Delta^{n-1}[0,t_n]} odB^{(i_1,\dots,i_{n-1})} \right) \circ dB^{i_n}_{t_n}.$$

$\square$

Proposition 1.1 *For $0 < s < t$,*

$$S(B)_t = S(B)_s \left(1 + \sum_{k=1}^{+\infty} \sum_{I=(i_1,\dots i_k)} X_{i_1}\dots X_{i_k} \int_{\Delta^k[s,t]} odB^I \right).$$

Proof. We have, thanks to the previous lemma,

$$S(B)_s \left(1 + \sum_{k=1}^{+\infty} \sum_I X_{i_1}\dots X_{i_k} \int_{\Delta^k[s,t]} odB^I \right)$$

$$= 1 + \sum_{k,k'=1}^{+\infty} \sum_{I,I'} X_{i_1}\dots X_{i_k} X_{i'_1}\dots X_{i'_{k'}} \int_{\Delta^k[s,t]} odB^I \int_{\Delta^{k'}[0,s]} odB^{I'}$$

$$= 1 + \sum_{k=1}^{+\infty} \sum_I X_{i_1}\dots X_{i_k} \int_{\Delta^k[0,t]} odB^I$$

$$= S(B)_t.$$

$\square$

Remark 1.3 *Observe that if $I \in \{1,\dots,d\}^k$ is a word with length k then for any $0 < s < t$:*

(1) $\int_{\Delta^k[s,t]} odB^I$ is independent from $(B_u)_{u\le s}$;
(2)

$$\int_{\Delta^k[s,t]} odB^I =^{law} \int_{\Delta^k[0,t-s]} odB^I.$$

Therefore, we can roughly conclude that:

(1) *for each $s \ge 0$, the process $\left(S(B)_s^{-1} S(B)_{t+s} \right)_{t\ge 0}$ is independent of the process $(S(B)_u)_{0\le u\le s}$;*

(2) *for each $s \geq 0$, the processes $(S(B)_s^{-1} S(B)_{t+s})_{t\geq 0}$ and $(S(B)_t)_{t\geq 0}$ are identical in law.*

By using the relation between Stratonovitch's and Itô's integral (see Appendix A), it is possible to give a formula for the signature of a Brownian motion which only involves Itô's iterated integrals.

Proposition 1.2 *We have*

$$S(B)_t = 1 + \sum_{k=1}^{+\infty} \sum_{I \in \{0,1,\dots,d\}^k} X_{i_1} \dots X_{i_k} \int_{\Delta^k[0,t]} dB^I, \ t \geq 0,$$

where we used the following notations:

(1)

$$X_0 = \frac{1}{2} \sum_{i=1}^{d} X_i^2, \ B_t^0 = t;$$

(2)

$$\int_{\Delta^k[0,t]} dB^I = \int_{0 \leq t_1 \leq \dots \leq t_k \leq t} dB_{t_1}^{i_1} \dots dB_{t_k}^{i_k}.$$

Proof. Let $I = (i_1, \dots, i_k) \in \{1, \dots, d\}^k$. From the definition of Stratonovitch's integrals, we have

$$\int_{\Delta^k[0,t]} \circ dB^I = \int_0^t \left(\int_{\Delta^{k-1}[0,t_k]} \circ dB^{i_1,\dots,i_{k-1}} \right) dB_{t_k}^{i_k}$$

$$+ \frac{1}{2} \tau_{i_{k-1},i_k} \int_0^t \left(\int_{\Delta^{k-2}[0,t_{k-1}]} \circ dB^{i_1,\dots,i_{k-2}} \right) dt_{k-1},$$

where

$$\tau_{i_{k-1},i_k} = 0 \text{ if } i_{k-1} \neq i_k$$
$$= 1 \text{ if } i_{k-1} = i_k.$$

Consider now the smallest set $\mathcal{I}$ of words which satisfies the following properties:

(1) $I \in \mathcal{I}$;
(2) if $J = (j_1, \dots, j_l) \in \mathcal{I}$ and if $j_m = j_{m+1} \neq 0$ for some $1 \leq m \leq l - 1$, then $(j_1, \dots, j_{m-1}, 0, j_{m+2}, \dots, j_l) \in \mathcal{I}$.

By iterating the previous formula, we get

$$\int_{\Delta^k[0,t]} \circ dB^I = \sum_{J \in \mathcal{I}} \frac{1}{2^{k-|J|}} \int_{\Delta^{|J|}[0,t]} dB^J,$$

where $|J|$ denotes the length of the word J. The expected result follows readily.   □

Remark 1.4 *Observe that if we write equation (1.4) in Itô's form, we get*

$$S(B)_t = 1 + \frac{1}{2}\int_0^t S(B)_s \left(\sum_{i=1}^d X_i^2\right) ds + \sum_{i=1}^d \int_0^t S(B)_s X_i dB_s^i,$$

which explains intuitively formula of Proposition 1.2.

1.3   The Chen-Strichartz development formula

This section is devoted to the proof of the Chen-Strichartz development formula. The formula we give is actually a restatement of a result of [Chen (1957)] and [Strichartz (1987)], and can be seen as a deep generalization of the Baker-Campbell-Hausdorff formula (see Appendix B).

     The Chen-Strichartz formula is an explicit formula for $\ln S(B)_t$. In particular, it appears that $\ln S(B)_t$ is a Lie element, a result which is far from being obvious at a first look. As it is illustrated in this book, the geometric consequences of this development are rather deep. Before we go into the heart of this formula, let us first try to understand a simple case: the commutative case.

     We denote $\mathfrak{S}_k$ the group of the permutations of the index set $\{1, ..., k\}$ and if $\sigma \in \mathfrak{S}_k$, we denote for a word $I = (i_1, ..., i_k)$, $\sigma \cdot I$ the word $(i_{\sigma(1)}, ..., i_{\sigma(k)})$. Now, let us observe that if $X_1, ..., X_d$ were commuting, we would have

$$S(B)_t = \exp\left(\sum_{i=1}^d X_i B_t^i\right).$$

Indeed in that case, by symmetrization, we get

$$S(B)_t = 1 + \sum_{k=1}^{+\infty} \sum_{I=(i_1,...,i_k)} X_{i_1}...X_{i_k}\left(\frac{1}{k!}\sum_{\sigma \in \mathfrak{S}_k}\int_{\Delta^k[0,t]} \circ dB^{\sigma \cdot I}\right).$$

Now observe that

$$\sum_{\sigma \in \mathfrak{S}_k} \int_{\Delta^k[0,t]} \circ dB^{\sigma \cdot I} = B_t^{i_1}...B_t^{i_k},$$

which implies,

$$S(B)_t = 1 + \sum_{k=1}^{+\infty} \frac{1}{k!} \sum_{I=(i_1,...,i_k)} X_{i_1}...X_{i_k} B_t^{i_1}...B_t^{i_k} = \exp\left(\sum_{i=1}^{d} X_i B_t^i\right).$$

Of course, in the general case, this formula does not hold anymore. But, we still have a nice formula for $\ln S(B)_t$ which involves iterated functionals of the commutators $X_i X_j - X_j X_i$.

We define the bracket between two elements U and V of $\mathbb{R}[[X_1, ..., X_d]]$ by

$$[U, V] = UV - VU,$$

and it is easily checked that this bracket endows $\mathbb{R}[[X_1, ..., X_d]]$ with a Lie algebra structure. If $I = (i_1, ..., i_k) \in \{1, ..., d\}^k$ is a word, we denote by X_I the commutator defined by

$$X_I = [X_{i_1}, [X_{i_2}, ..., [X_{i_{k-1}}, X_{i_k}]...].$$

If $\sigma \in \mathfrak{S}_k$, we denote $e(\sigma)$ the cardinality of the set

$$\{j \in \{1, ..., k-1\}, \sigma(j) > \sigma(j+1)\}.$$

Theorem 1.1   *We have*

$$S(B)_t = \exp\left(\sum_{k \geq 1} \sum_{I=(i_1,...,i_k)} \Lambda_I(B)_t X_I\right), \quad t \geq 0,$$

where:

$$\Lambda_I(B)_t = \sum_{\sigma \in \mathfrak{S}_k} \frac{(-1)^{e(\sigma)}}{k^2 \binom{k-1}{e(\sigma)}} \int_{\Delta^k[0,t]} \circ dB^{\sigma^{-1} \cdot I}.$$

Proof.   We shall proceed in several steps.

Step 1. First, we write

$$S(B)_t = 1 + \sum_{k=1}^{+\infty} \int_{\Delta^k[0,t]} \circ dw_{s_1} ... \circ dw_{s_k},$$

where we used the notation

$$dw = \sum_{i=1}^{d} X_i dB^i.$$

Now we have,

$$S(B)_t = \exp\left(\ln S(B)_t\right) = \exp\left(\sum_{k=1}^{+\infty} \frac{(-1)^{k-1}}{k}(S(B)_t - 1)^k\right).$$

Therefore we get,

$$S(B)_t = \exp Z_t,$$

with

$$Z_t = \sum_{k=1}^{+\infty} \frac{(-1)^{k-1}}{k}\left(\sum_{n=1}^{+\infty}\int_{\Delta^n[0,t]} \circ dw_{s_1}\ldots \circ dw_{s_n}\right)^k. \qquad (1.6)$$

For each positive integer r, consider all ways of writing

$$r = p_1 + \ldots + p_m, \quad m = 1, \ldots, r,$$

for p_j positive integers, and set $q_0 = 0$ and $q_j = p_1 + \ldots + p_j$, for $j \geq 1$. We can now expand out

$$\sum_{k=1}^{+\infty} \frac{(-1)^{k-1}}{k}\left(\sum_{n=1}^{+\infty}\int_{\Delta^n[0,t]} \circ dw_{s_1}\ldots \circ dw_{s_n}\right)^k,$$

to obtain, thanks to Lemma 1.1,

$$Z_t = \sum_{r=1}^{+\infty}\sum_{m=1}^{r}\sum_{p_j} \frac{(-1)^{m-1}}{m}\int \circ dw_{s_1}\ldots \circ dw_{s_r},$$

where the integral is taken over the region given by the inequalities

$$\begin{aligned} &0 < s_1 < \ldots < s_{q_1} < t, \\ &0 < s_{q_1+1} < \ldots < s_{q_2} < t \\ &\ldots \\ &0 < s_{q_{m-1}+1} < \ldots < s_{q_m} < t. \end{aligned} \qquad (1.7)$$

Step 2. By applying now the generalized Baker-Campbell-Hausdorff formula (B.4) of Appendix B, we obtain

$$Z_t = \sum_{r=1}^{+\infty} \sum_{m=1}^{r} \sum_{p_j} \frac{(-1)^{m-1}}{mr} \int [...[\circ d\omega_{s_1}, \circ d\omega_{s_2}]...], \circ d\omega_{s_r}], \qquad (1.8)$$

where the integral is taken over the same region. The domain determined by the inequalities (1.7) can be written as the union of simplices obtained from the simplex $\Delta^r[0, t]$ by permuting the variables, actually

$$\int [...[\circ d\omega_{s_1}, \circ d\omega_{s_2}]...], \circ d\omega_{s_r}] = \sum \int_{\Delta^r[0,t]} [...[\circ d\omega_{s_{\sigma(1)}}, \circ d\omega_{s_{\sigma(2)}}]...], \circ d\omega_{s_{\sigma(r)}}],$$

where the inner sum is taken over the permutations $\sigma \in \mathfrak{S}_r$ that satisfy

$$\sigma(q_j + 1) < \sigma(q_j + 2) < ... < \sigma(q_{j+1}), \ j = 0, \cdots, m-1.$$

Therefore, by regrouping the terms in (1.8), we obtain that Z_t is equal to

$$\sum_{r=1}^{+\infty} \sum_{\sigma \in \mathfrak{S}_r} \sum_{m=1}^{r} \frac{(-1)^{m-1}}{mr} d(r, m, \sigma) \int_{\Delta^r[0,t]} [...[\circ d\omega_{s_{\sigma(1)}}, \circ d\omega_{s_{\sigma(2)}}]...], \circ d\omega_{s_{\sigma(r)}}],$$

where $d(r, m, \sigma)$ is the number of ways of choosing positive integers $p_1, ..., p_m$ with $p_1 + ... + p_m = r$ satisfying $\sigma(q_j + 1) < ... < \sigma(q_{j+1})$, $j = 0, ..., m-1$.

Step 3. We claim now that

$$\sum_{m=1}^{r} \frac{(-1)^{m-1}}{mr} d(r, m, \sigma) = \frac{(-1)^{e(\sigma)}}{r^2 \binom{r-1}{e(\sigma)}}.$$

Indeed, a straightforward combinatorial argument shows that

$$d(r, m, \sigma) = \binom{r - e(\sigma) - 1}{m - e(\sigma) - 1},$$

so that we need to sum

$$\sum_{m=e(\sigma)+1}^{r} \frac{(-1)^{m-1}}{mr} \binom{r - e(\sigma) - 1}{m - e(\sigma) - 1}.$$

But let us observe that for $n \geq 0$ and $k > 0$,

$$\sum_{j=0}^{n} \frac{(-1)^j}{k+j} \binom{n}{j} = \sum_{j=0}^{n} (-1)^j \binom{n}{j} \int_0^1 x^{j+k-1} dx$$

$$= \int_0^1 (1-x)^n \, x^{k-1} dx$$

$$= \frac{(k-1)! \, n!}{(k+n)!}.$$

Therefore, by setting $k = e(\sigma) + 1$, $m = k + j$, and $n = r - k$ we obtain

$$\sum_{m=1}^{r} \frac{(-1)^{m-1}}{mr} d(r, m, \sigma) = \frac{(-1)^{e(\sigma)}}{r^2 \binom{r-1}{e(\sigma)}}.$$

Step 4. Putting things together, we get therefore

$$Z_t = \sum_{r=1}^{+\infty} \sum_{\sigma \in \mathfrak{S}_r} \frac{(-1)^{e(\sigma)}}{r^2 \binom{r-1}{e(\sigma)}} \int_{\Delta^r[0,t]} [\dots[\circ dw_{s_{\sigma(1)}}, \circ dw_{s_{\sigma(2)}}]\dots], \circ dw_{s_{\sigma(r)}}].$$

Now, observe that

$$[\circ dw_{s_{\sigma(1)}}, [\dots, [\circ dw_{s_{\sigma(r-1)}}, \circ dw_{s_{\sigma(r)}}]\dots]$$
$$= (-1)^{r-1}[\dots[\circ dw_{s_{\sigma(r)}}, \circ dw_{s_{\sigma(r-1)}}]\dots], \circ dw_{s_{\sigma(1)}}],$$

and that if for $\sigma \in \mathfrak{S}_r$, σ^* denotes the permutation defined by $\sigma^*(k) = \sigma(r + 1 - k)$, then $e(\sigma^*) = r - 1 - e(\sigma)$. Therefore, we also have

$$Z_t = \sum_{r=1}^{+\infty} \sum_{\sigma \in \mathfrak{S}_r} \frac{(-1)^{e(\sigma)}}{r^2 \binom{r-1}{e(\sigma)}} \int_{\Delta^r[0,t]} [\circ dw_{s_{\sigma(1)}}, [\dots, [\circ dw_{s_{\sigma(r-1)}}, \circ dw_{s_{\sigma(r)}}]\dots].$$

By expanding out

$$\sum_{\sigma \in \mathfrak{S}_r} \frac{(-1)^{e(\sigma)}}{r^2 \binom{r-1}{e(\sigma)}} \int_{\Delta^r[0,t]} [\circ dw_{s_{\sigma(1)}}, [\dots, [\circ dw_{s_{\sigma(r-1)}}, \circ dw_{s_{\sigma(r)}}]\dots]$$

into

$$\sum_{\sigma \in \mathfrak{S}_r} \frac{(-1)^{e(\sigma)}}{r^2 \binom{r-1}{e(\sigma)}} \sum_{I=(i_1,\ldots,i_r)} X_I \int_{\Delta^k[0,t]} \circ dB^{\sigma^{-1} \cdot I},$$

we obtain finally the claimed formula. $\qquad\qquad\qquad\qquad\qquad\square$

Remark 1.5 *Observe that the first terms in the Chen-Strichartz formula are:*

(1)

$$\sum_{I=(i_1)} \Lambda_I(B)_t X_I = \sum_{k=1}^{d} B_t^i X_i;$$

(2)

$$\sum_{I=(i_1,i_2)} \Lambda_I(B)_t X_I = \frac{1}{2} \sum_{1 \le i < j \le d} [X_i, X_j] \int_0^t B_s^i \circ dB_s^j - B_s^j \circ dB_s^i;$$

it is interesting to note the above Stratonovitch integrals are also Itô integrals, that is

$$\int_0^t B_s^i \circ dB_s^j - B_s^j \circ dB_s^i = \int_0^t B_s^i dB_s^j - B_s^j dB_s^i.$$

Remark 1.6 *Actually, the Chen-Strichartz formula holds for the signature of any semimartingale: this is indeed a pathwise result.*

Remark 1.7 *The formal development for the action on functions of the stochastic flow Φ_t^* associated with a stochastic differential equation of the type (1.3) reads therefore*

$$\Phi_t^* = \exp\left(\sum_{k \ge 1} \sum_{I=(i_1,\ldots,i_k)} \Lambda_I(B)_t V_I \right).$$

It is also possible to obtain a formal development for the action of Φ_t on smooth tensor fields. Indeed an iteration of the formula given in the Proposition A.6 of Appendix A leads, due to the Lie algebra homomorphism property of the Lie derivative, to

$$\Phi_t^* = \exp\left(\sum_{k \ge 1} \sum_{I=(i_1,\ldots,i_k)} \Lambda_I(B)_t \mathcal{L}_{V_I} \right).$$

1.4   Expectation of the signature of a Brownian motion

It is interesting to note that it is possible to derive in a purely algebraic manner the semigroup P_t associated with the solution of a stochastic differential equation driven by Brownian motions .

Definition 1.3   The element of $\mathbb{R}[[X_1, ..., X_d]]$ defined by

$$P_t = 1 + \sum_{k=1}^{+\infty} \sum_{I=(i_1,...i_k)} X_{i_1}...X_{i_k} \mathbb{E}\left(\int_{\Delta^k[0,t]} \circ dB^I\right), \ t \geq 0,$$

is called the expectation of the signature of the Brownian motion $(B_t)_{t \geq 0}$.

Proposition 1.3   *We have*

$$P_t = \exp\left(\frac{1}{2}t\sum_{i=1}^{d} X_i^2\right), \ t \geq 0.$$

Proof.   By using proposition (1.2), we get

$$P_t = 1 + \sum_{k=1}^{+\infty} \sum_{I \in \{0,1,...,d\}^k} X_{i_1}...X_{i_k} \mathbb{E}\left(\int_{\Delta^k[0,t]} dB^I\right), \ t \geq 0,$$

where:

$$X_0 = \frac{1}{2}\sum_{i=1}^{d} X_i^2, \text{ and } B_t^0 = t.$$

In the previous sum, the only terms whose expectation does not vanish are the terms

$$\int_{\Delta^k[0,t]} dB^I$$

where I is a word which contains only 0. Therefore,

$$P_t = 1 + \sum_{k=1}^{+\infty} X_0^k \int_{\Delta^k[0,t]} dt_1...dt_k.$$

Since

$$\int_{\Delta^k[0,t]} dt_1...dt_k = \frac{1}{k!}\int_{[0,t]^k} dt_1...dt_k = \frac{t^k}{k!},$$

we get

$$P_t = 1 + \sum_{k=1}^{+\infty} \frac{t^k}{k!} X_0^k = e^{tX_0}.$$

$\square$

Remark 1.8 *If we think* $(S(B)_t)_{t\geq 0}$ *as the solution of the formal stochastic differential equation*

$$S(B)_t = 1 + \sum_{i=1}^{d} \int_0^t S(B)_s X_i \circ dB_s^i, \qquad (1.9)$$

and P_t *as* $\mathbb{E}(S(B)_t)$, *then the above formula is rather intuitive. Indeed, by writing the Itô's form of (1.9), and by taking the expectation, we obtain the equation*

$$P_t = 1 + \int_0^t P_s \left(\frac{1}{2} \sum_{i=1}^{d} X_i^2 \right) ds,$$

which directly implies

$$P_t = \exp\left(\frac{1}{2} t \sum_{i=1}^{d} X_i^2 \right).$$

Remark 1.9 *In the commutative case, we have*

$$S(B)_t = \exp\left(\sum_{i=1}^{d} X_i B_t^i \right),$$

and the formula for P_t *reduces to the well-known Laplace transform formula*

$$\mathbb{E}\left(\exp\left(\sum_{i=1}^{d} X_i B_t^i \right) \right) = \exp\left(\frac{1}{2} t \sum_{i=1}^{d} X_i^2 \right).$$

We stress the fact that the last formula only holds in the commutative case.

Observe that the semigroup property of P_t, that is

$$P_{t+s} = P_t P_s,$$

could have been directly derived from the identity

$$S(B)_{t+s} = S(B)_t \left(1 + \sum_{k=1}^{+\infty} \sum_{I=(i_1,\dots i_k)} X_{i_1} \dots X_{i_k} \int_{\Delta^k[t,t+s]} \circ dB^I \right).$$

Indeed, since $\int_{\Delta^k[t,t+s]} \circ dB^I$ is independent of $(B_u)_{u \leq t}$, we deduce

$$P_{t+s} = P_t \left(1 + \sum_{k=1}^{+\infty} \sum_{I=(i_1,\dots i_k)} X_{i_1} \dots X_{i_k} \mathbb{E} \left(\int_{\Delta^k[t,t+s]} \circ dB^I \right) \right).$$

Now observe that, due to the stationarity of the increments of a Brownian motion,

$$\mathbb{E} \left(\int_{\Delta^k[t,t+s]} \circ dB^I \right) = \mathbb{E} \left(\int_{\Delta^k[0,s]} \circ dB^I \right),$$

so that

$$P_{t+s} = P_t \left(1 + \sum_{k=1}^{+\infty} \sum_{I=(i_1,\dots i_k)} X_{i_1} \dots X_{i_k} \mathbb{E} \left(\int_{\Delta^k[0,s]} \circ dB^I \right) \right) = P_t P_s.$$

We have already pointed out that the signature is a universal object in the theory of stochastic flows, so let us see the analytic counterpart of the purely algebraic formula

$$1 + \sum_{k=1}^{+\infty} \sum_{I=(i_1,\dots i_k)} X_{i_1} \dots X_{i_k} \mathbb{E} \left(\int_{\Delta^k[0,t]} \circ dB^I \right) = \exp \left(\frac{1}{2} t \sum_{i=1}^{d} X_i^2 \right).$$

In the first section, we have seen that for the action on smooth functions of the stochastic flow Φ associated with the stochastic differential equation

$$X_t^{x_0} = x_0 + \sum_{i=1}^{d} \int_0^t V_i(X_s^{x_0}) \circ dB_s^i,$$

we had formally

$$\Phi_t^* = \mathbf{Id} + \sum_{k=1}^{+\infty} \sum_{I=(i_1,\dots i_k)} V_{i_1} \dots V_{i_k} \int_{\Delta^k[0,t]} \circ dB^I.$$

Therefore,

$$\mathbb{E} \left(\Phi_t^* \right) = \mathbf{Id} + \sum_{k=1}^{+\infty} \sum_{I=(i_1,\dots i_k)} V_{i_1} \dots V_{i_k} \mathbb{E} \left(\int_{\Delta^k[0,t]} \circ dB^I \right) = e^{\frac{1}{2} t \sum_{i=1}^{d} V_i^2}.$$

By coming back to the definition of Φ_t^* we deduce that if $f : \mathbb{R}^n \to \mathbb{R}$ is a smooth function,

$$\mathbb{E}\left(f(X_t^{x_0})\right) = \left(e^{\frac{1}{2}t\sum_{i=1}^d V_i^2} f\right)(x_0),$$

which exactly says that $(X_t^{x_0})_{t\geq 0}$ is a Markov process with generator $e^{\frac{1}{2}t\sum_{i=1}^d V_i^2}$. In the same way, by using the formal development of Φ_t on smooth tensor fields which reads

$$\Phi_t^* = \mathrm{Id} + \sum_{k=1}^{+\infty} \sum_{I=(i_1,\dots i_k)} \mathcal{L}_{V_{i_1}}\dots\mathcal{L}_{V_{i_k}} \int_{\Delta^k[0,t]} \circ dB^I,$$

where $\mathcal{L}$ denotes the Lie derivative, we obtain that if K is a smooth tensor field on $\mathbb{R}^n$,

$$\mathbb{E}\left[(\Phi_t^* K)(x_0)\right] = \left(e^{\frac{1}{2}t\sum_{i=1}^d \mathcal{L}_{V_i}^2} K\right)(x_0).$$

Of course, all this is only formal, but should convince the reader of the relevance of the formal calculus on the signature.

1.5 Expectation of the signature of other processes

As already observed, the notion of signature can be defined for other processes than Brownian motions and there is a corresponding notion of expectation for the signature. Let us for instance mention the example of the signature of a fractional Brownian motion. A d-dimensional fractional Brownian motion with Hurst parameter $H > \frac{1}{2}$ is a Gaussian process

$$B_t = (B_t^1, \dots, B_t^d), \ t \geq 0,$$

where $B^1, \dots, B^d$ are d independent centered Gaussian processes with covariance function

$$R(t,s) = \frac{1}{2}\left(s^{2H} + t^{2H} - |t - s|^{2H}\right).$$

It can be shown that such a process admits a continuous version whose paths are locally p-Hölder for $p < H$. Therefore, if $H > \frac{1}{2}$, the integrals

$$\int_{\Delta^k[0,t]} dB^I$$

can be understood in the sense of Young's integration; see [Young (1936)] and [Zähle (1998)]. We define then the signature of $(B_t)_{t \geq 0}$ by

$$S(B)_t = 1 + \sum_{k=1}^{+\infty} \sum_{I=(i_1,\dots i_k)} X_{i_1}\dots X_{i_k} \int_{\Delta^k[0,t]} dB^I, \ t \geq 0,$$

and associate with $S(B)$ the family of operators

$$P_t = 1 + \sum_{k=1}^{+\infty} \sum_{I=(i_1,\dots i_k)} X_{i_1}\dots X_{i_k} \mathbb{E}\left(\int_{\Delta^k[0,t]} dB^I\right), \ t \geq 0.$$

The increments of $(B_t)_{t \geq 0}$ are not independent (they are however stationary), and $(P_t)_{t \geq 0}$ is therefore not a semigroup. Nevertheless, as shown in [Baudoin and Coutin (2004)], when $t \to 0$,

$$P_t = 1 + \frac{1}{2}t^{2H}\left(\sum_{i=1}^{d} X_i^2\right) + t^{4H} \sum_{i,j,k,l=1}^{d} a_{i,j,k,l} X_i X_j X_k X_l + O(t^{6H}),$$

where,

$$a_{i,j,k,l} = \frac{1}{2}\delta_{k,l}\delta_{j,i}\left[\frac{1}{4} - 2H\beta(2H, 2H-1)\right] + \frac{1}{2}\delta_{i,k}\delta_{j,l}\frac{2H-1}{4H(4H-1)}$$
$$+ \frac{H(2H-1)}{8}\delta_{j,k}\delta_{i,l}\left[\beta(2H, 2H-2) + \frac{1}{4H-1} - \frac{1}{2H-1}\right],$$

with $\beta(x,y) = \int_0^1 t^{x-1}(1-t)^{y-1}dt$, and $\delta_{i,j}$ is the Kronecker's symbol. A development for P_t which leads to development in small times of expressions of the type $\mathbb{E}\left(f(X_t^{x_0})\right)$, where $(X_t^{x_0})_{t \geq 0}$ denotes the solution of the equation

$$X_t^{x_0} = x_0 + \sum_{i=1}^{d} \int_0^t V_i(X_s^{x_0})dB_s^i, \ t \geq 0,$$

which is understood in Young's sense (see [Nualart and Rascanu (2002)] for theorems concerning the existence and the uniqueness for the solution of such an equation).

Observe that when $H \to \frac{1}{2}$, then the above development tends to

$$P_t = 1 + \frac{1}{2}t\left(\sum_{i}^{d} X_i^2\right) + \frac{1}{8}t^2\left(\sum_{i}^{d} X_i^2\right)^2 + O(t^3),$$

which is the development of the P_t corresponding to the Brownian motion.

Also observe that the fourth order operator

$$\sum_{i,j,k,l=1}^{d} a_{i,j,k,l} X_i X_j X_k X_l$$

can not be simply expressed from

$$\sum_{i=1}^{d} X_i^2.$$

Such a discussion can obviously be generalized to any stochastic differential equation driven by Gaussian processes whose paths are more than $\frac{1}{2}$ locally Hölder continuous and this is actually an interesting open question to decide what is the smallest sub-algebra of $\mathbb{R}[[X_1, ..., X_d]]$ that contains P_t.

Let us finally mention another type of processes for which the expectation of the signature can be explicitly computed. Let us consider the process

$$Z_t = B_{\sigma t}, \quad t \geq 0,$$

where $(B_t)_{t \geq 0}$ is a d-dimensional standard Brownian motion and σ a non negative random variable independent of $(B_t)_{t \geq 0}$ which satisfies $\mathbb{E}\left(\sigma^k\right) < +\infty$, $k \geq 0$. In that case, the expectation of the signature of $(Z_t)_{t \geq 0}$ is easily seen to be given by

$$P_t = \sum_{k=0}^{+\infty} \frac{1}{2^k k!} \mathbb{E}\left(\sigma^k\right) t^k \left(\sum_{i=1}^{d} X_i^2\right)^k,$$

and observe that, like in the Brownian case, the smallest algebra containing P_t is given by $\mathbb{R}\left[\sum_{i=1}^{d} X_i^2\right]$. For instance, by taking for σ an exponential law with parameter 1, that is

$$\mathbb{P}(\sigma \in dx) = e^{-x} 1_{\mathbb{R}_{\geq 0}}(x),$$

we get

$$P_t = \frac{1}{1 - \frac{1}{2}t \left(\sum_{i=1}^{d} X_i^2\right)}.$$

Chapter 2

Stochastic Differential Equations and Carnot Groups

Let us consider a stochastic differential equation

$$X_t^{x_0} = x_0 + \sum_{i=1}^{d} \int_0^t V_i(X_s^{x_0}) \circ dB_s^i, \ t \geq 0, \tag{2.1}$$

where $x_0 \in \mathbb{R}^n$, $V_1, ..., V_d$ are C^∞ bounded vector fields on $\mathbb{R}^n$ and $(B_t)_{t \geq 0}$ is a d-dimensional standard Brownian motion. Since $(X_t^{x_0})_{t \geq 0}$ is a strong solution of (2.1), we know from the general theory of stochastic differential equations that $(X_t^{x_0})_{t \geq 0}$ is a predictable functional of $(B_t)_{t \geq 0}$ (see Appendix A).

In this chapter, we would like to better understand this pathwise representation. For this, the best tool is certainly the Chen-Strichartz formula which has been proved in the first chapter. Indeed, if we denote $(\Phi_t)_{t \geq 0}$ the stochastic flow associated with equation (2.1), then the Chen-Strichartz formula shows that for the action of Φ on smooth functions

$$\Phi_t^* = \exp \left(\sum_{k \geq 1} \sum_{I=(i_1,...,i_k)} \Lambda_I(B)_t V_I \right),$$

where

$$V_I = [V_{i_1}, [V_{i_2}, ..., [V_{i_{k-1}}, V_{i_k}]...],$$

and

$$\Lambda_I(B)_t = \sum_{\sigma \in \mathfrak{S}_k} \frac{(-1)^{e(\sigma)}}{k^2 \binom{k-1}{e(\sigma)}} \int_{\Delta^k[0,t]} \circ dB^{\sigma^{-1} \cdot I}.$$

21

Even though this is only a formal development, it clearly shows how the dependance between B and X^{x_0} is related to the structure of the Lie algebra $\mathcal{L}$ generated by the vector fields V_i's. If we want to understand more deeply how the properties of this Lie algebra determine the geometry of X^{x_0}, it is wiser to begin with the simplest cases. In a way, the most simple Lie algebras are the nilpotent ones. In that case, that is if $\mathcal{L}$ is nilpotent, then the sum

$$\sum_{k \geq 1} \sum_{I=(i_1,\ldots,i_k)} \Lambda_I(B)_t V_I$$

is actually finite and we shall show that the solutions of equation (2.1) can be represented from the lift of the Brownian motion $(B_t)_{t \geq 0}$ in a graded free nilpotent Lie group with dilations. These groups called the free Carnot groups are introduced and their geometries are discussed.

When the Lie algebra $\mathcal{L}$ is not nilpotent, this representation does not hold anymore but provides a good approximation for X^{x_0} in small times. We conclude the chapter with an introduction to the rough paths theory of [Lyons (1998)], in which Carnot groups also play a fundamental role.

2.1 The commutative case

In this section, we shall assume that the Lie algebra $\mathcal{L} = \mathbf{Lie}(V_1, \ldots, V_d)$ is commutative, i.e. that $[V_i, V_j] = 0$ for $1 \leq i, j \leq d$. This is therefore the simplest possible case: it has been first studied by [Doss (1977)] and [Süssmann (1978)]. The main theorem is the following:

Theorem 2.1 *There exists a smooth map*

$$F : \mathbb{R}^n \times \mathbb{R}^d \to \mathbb{R}^n$$

such that, for $x_0 \in \mathbb{R}^n$, the solution $(X_t^{x_0})_{t \geq 0}$ of the stochastic differential equation (2.1) can be written

$$X_t^{x_0} = F(x_0, B_t), \ t \geq 0.$$

Proof. For $i = 1, \ldots, d$, let us denote by $(e^{tV_i})_{t \in \mathbb{R}}$ the flow associated with the ordinary differential equation

$$\frac{dx}{dt} = V_i(x_t).$$

Observe that since the vector fields V_i's are commuting, these flows are also commuting (see Appendix B). We set now for $(x, y) \in \mathbb{R}^n \times \mathbb{R}^d$,

$$F(x, y) = \left(e^{y_1 V_1} \circ \dots \circ e^{y_d V_d} \right)(x).$$

By applying Itô's formula, we easily see that the process $\left(e^{B_t^d V_d} x_0 \right)_{t \geq 0}$ is solution of the SDE

$$d \left(e^{B_t^d V_d}(x_0) \right) = V_d \left(e^{B_t^d V_d}(x_0) \right) \circ dB_t^d.$$

A new application of Itô's formula shows now that, since V_d and V_{d-1} are commuting,

$$d \left(e^{B_t^{d-1} V_{d-1}} (e^{B_t^d V_d} x_0) \right) =$$

$$V_{d-1} \left(e^{B_t^{d-1} V_{d-1}} (e^{B_t^d V_d} x_0) \right) \circ dB_t^{d-1} + V_d \left(e^{B_t^{d-1} V_{d-1}} (e^{B_t^d V_d} x_0) \right) \circ dB_t^d.$$

We deduce hence, by an iterative application of Itô's formula that the process $(F(x_0, B_t))_{t \geq 0}$ satisfies

$$dF(x_0, B_t) = \sum_{i=1}^{d} V_i(F(x_0, B_t)) \circ dB_t^i.$$

Thus, by pathwise uniqueness for the stochastic differential equation (2.1), we conclude that

$$X_t^{x_0} = F(x_0, B_t), \ t \geq 0.$$

$\square$

Computing the function F which appears in the above theorem is not possible in all generality. Indeed, the proof has shown that the effective computation of F is equivalent to the explicit resolution of the d ordinary differential equations

$$\frac{dx_t}{dt} = V_i(x_t),$$

which is of course not an easy matter.

There is however a case of particular importance where it is possible to solve explicitly the stochastic differential equation (2.1): this is the case $n = d = 1$. Indeed, in that case, (2.1) can be written

$$dX_t = v(X_t) \circ dB_t,$$

which also reads in Itô's form

$$dX_t = \frac{1}{2}v(X_t)v'(X_t)dt + v(X_t)dB_t.$$

This equation is solved by

$$X_t = g(B_t),$$

where g solves

$$g' = v \circ g.$$

2.2   Two-step nilpotent SDE's

In this section we introduce the free 2-step Carnot group over $\mathbb{R}^d$, study its geometry and show that it is universal in the study of 2-step nilpotent diffusions. Almost all what follows will be generalized in the next section. Nevertheless, we believe that the understanding of the geometry of the free 2-step Carnot group is much more easy to get than the geometry of general Carnot groups and however contains the most important ideas of sub-Riemannian geometry. Let $d \geq 2$ and denote $\mathcal{AS}_d$ the space of $d \times d$ skew-symmetric matrices. We consider the group $\mathbb{G}_2(\mathbb{R}^d)$ defined in the following way

$$\mathbb{G}_2(\mathbb{R}^d) = (\mathbb{R}^d \times \mathcal{AS}_d, \circledast)$$

where $\circledast$ is the group law defined by

$$(\alpha_1, \omega_1) \circledast (\alpha_2, \omega_2) = (\alpha_1 + \alpha_2, \omega_1 + \omega_2 + \frac{1}{2}\alpha_1 \wedge \alpha_2).$$

Here we use the following notation; if $\alpha_1, \alpha_2 \in \mathbb{R}^d$, then $\alpha_1 \wedge \alpha_2$ denotes the skew-symmetric matrix $\left(\alpha_1^i \alpha_2^j - \alpha_1^j \alpha_2^i\right)_{i,j}$. Consider now the vector fields

$$D_i(x) = \frac{\partial}{\partial x^i} + \frac{1}{2}\sum_{j<i} x^j \frac{\partial}{\partial x^{j,i}} - \frac{1}{2}\sum_{j>i} x^j \frac{\partial}{\partial x^{i,j}}, \ 1 \leq i \leq d,$$

defined on $\mathbb{R}^d \times \mathcal{AS}_d$. It is easy to check that:

(1) For $x \in \mathbb{R}^d \times \mathcal{AS}_d$,

$$[D_i, D_j](x) = \frac{\partial}{\partial x^{i,j}}, \ 1 \leq i < j \leq d;$$

(2) For $x \in \mathbb{R}^d \times \mathcal{AS}_d$,

$$[[D_i, D_j], D_k](x) = 0, \ 1 \leq i, j, k \leq d;$$

(3) The vector fields

$$(D_i, [D_j, D_k])_{1 \leq i \leq d, 1 \leq j < k \leq d}$$

are invariant with respect to the left action of $\mathbb{G}_2(\mathbb{R}^d)$ on itself and form a basis of the Lie algebra $\mathfrak{g}_2(\mathbb{R}^d)$ of $\mathbb{G}_2(\mathbb{R}^d)$.

It follows that $\mathbb{G}_2(\mathbb{R}^d)$ is a $\frac{d(d+1)}{2}$-dimensional step-two nilpotent Lie group whose Lie algebra can be written

$$\mathfrak{g}_2(\mathbb{R}^d) = \mathbb{R}^d \oplus [\mathbb{R}^d, \mathbb{R}^d].$$

Notice, moreover, that the scaling

$$c \cdot (\alpha, \omega) = (c\alpha, c^2\omega) \tag{2.2}$$

defines an automorphism of the group $\mathbb{G}_2(\mathbb{R}^d)$.

Definition 2.1 The group $\mathbb{G}_2(\mathbb{R}^d)$ is called the free two-step Carnot group over $\mathbb{R}^d$.

Example 2.1

(1) The Heisenberg group $\mathbb{H}$ can be represented as the set of 3×3 matrices:

$$\begin{pmatrix} 1 & x & z \\ 0 & 1 & y \\ 0 & 0 & 1 \end{pmatrix}, \ x, y, z \in \mathbb{R}.$$

The Lie algebra of $\mathbb{H}$ is spanned by the matrices

$$D_1 = \begin{pmatrix} 0 & 1 & 0 \\ 0 & 0 & 0 \\ 0 & 0 & 0 \end{pmatrix}, \ D_2 = \begin{pmatrix} 0 & 0 & 0 \\ 0 & 0 & 1 \\ 0 & 0 & 0 \end{pmatrix} \text{ and } D_3 = \begin{pmatrix} 0 & 0 & 1 \\ 0 & 0 & 0 \\ 0 & 0 & 0 \end{pmatrix},$$

for which the following equalities hold

$$[D_1, D_2] = D_3, \ [D_1, D_3] = [D_2, D_3] = 0.$$

Thus

$$\mathfrak{h} \sim \mathbb{R}^2 \oplus [\mathbb{R}, \mathbb{R}],$$

and,

$$\mathbb{H} \sim \mathbb{G}_2(\mathbb{R}^2).$$

(2) Let us mention a pathwise point of view on the law of $\mathbb{G}_2(\mathbb{R}^2)$ which has been pointed to us by N. Victoir. If $x : [0, +\infty) \to \mathbb{R}^2$ is an absolutely continuous path, then for $0 < t_1 < t_2$ we denote

$$\Delta_{[t_1,t_2]} x = \left(x_{t_2}^1 - x_{t_1}^1, x_{t_2}^2 - x_{t_1}^2, S_{[t_1,t_2]} x \right),$$

where $S_{[t_1,t_2]} x$ is the area swept out by the vector $\overrightarrow{x_{t_1} x_t}$ during the time interval $[t_1, t_2]$. Then, it is easily checked that for $0 < t_1 < t_2 < t_3$,

$$\Delta_{[t_1,t_3]} x = \Delta_{[t_1,t_2]} x \circledast \Delta_{[t_2,t_3]} x,$$

where $\circledast$ is precisely the law of $\mathbb{G}_2(\mathbb{R}^2)$, i.e. for (x_1, y_1, z_1), $(x_2, y_2, z_2) \in \mathbb{R}^3$,

$$(x_1, y_1, z_1) \circledast (x_2, y_2, z_2) = \left(x_1 + x_2, y_1 + y_2, z_1 + z_2 + \frac{1}{2} (x_1 y_2 - x_2 y_1) \right).$$

Let now $(B_t)_{t \geq 0}$ be the Brownian motion considered in the equation (2.1). There exists a unique $\mathbb{G}_2(\mathbb{R}^d)$-valued semimartingale $(B_t^*)_{t \geq 0}$ such that $B_0^* = 0_{\mathbb{G}_2(\mathbb{R}^d)}$ and

$$dB_t^* = \sum_{i=1}^{d} D_i(B_t^*) \circ dB_s^i, \ 0 \leq t \leq T.$$

Definition 2.2   The process $(B_t^*)_{t \geq 0}$ is called the lift of the Brownian motion $(B_t)_{t \geq 0}$ in the group $\mathbb{G}_2(\mathbb{R}^d)$.

It is immediate to check that we have

$$B_t^* = \left(B_t, \frac{1}{2} \left(\int_0^t B_s^i \circ dB_s^j - B_s^j \circ dB_s^i \right)_{1 \leq i,j \leq d} \right), \ t \geq 0.$$

It is interesting to note the above Stratonovitch integrals are also Itô integrals, i.e.

$$\int_0^t B_s^i \circ dB_s^j - B_s^j \circ dB_s^i = \int_0^t B_s^i dB_s^j - B_s^j dB_s^i.$$

Also observe that we have the following scaling property, for every $c > 0$,

$$(B_{ct}^*)_{t \geq 0} =^{\text{law}} \left(\sqrt{c} \cdot B_t^* \right)_{t \geq 0}.$$

The process B^* can be seen as a diffusion process in $\mathbb{R}^d \times \mathcal{AS}_d$ (sometimes called the Gaveau diffusion, see [Gaveau (1977)] and [Malliavin (1997)]) whose generator is given by

$$\frac{1}{2} \sum_{i=1}^{d} \frac{\partial^2}{\partial(x^i)^2} + \frac{1}{2} \sum_{i<j} \left(x^i \frac{\partial}{\partial x^j} - x^j \frac{\partial}{\partial x^i} \right) \frac{\partial}{\partial x^{i,j}} + \frac{1}{8} \sum_{i<j} ((x^i)^2 + (x^j)^2) \frac{\partial^2}{\partial(x^{i,j})^2}.$$

Moreover, as shown by [Gaveau (1977)], the law of B^* is characterized by the following generalization of Lévy's area formula:

Lemma 2.1 *Let A be a $d \times d$ skew-symmetric matrix. Then, for $t > 0$,*

$$\mathbb{E} \left(e^{i \int_0^t (AB_s, dB_s)} \mid B_t = z \right) = \det \left(\frac{tA}{\sin tA} \right)^{\frac{1}{2}} \exp \left(\frac{I - tA \cot tA}{2t} z, z \right).$$

It is now time to say few words about the natural geometry of $\mathbb{G}_2(\mathbb{R}^d)$. First, we note that there is a natural scalar product g associated with the previous operator, precisely we define for $(\alpha, \omega), (\alpha', \omega') \in \mathfrak{g}_2(\mathbb{R}^d)$,

$$g((\alpha, \omega), (\alpha', \omega')) = \langle \alpha, \alpha' \rangle_{\mathbb{R}^d},$$

where $\langle, \rangle_{\mathbb{R}^d}$ denotes the usual scalar product on $\mathbb{R}^d$. This scalar product, only defined on $\mathfrak{g}_2(\mathbb{R}^d)$, i.e. on the tangent space to $\mathbb{G}_2(\mathbb{R}^d)$ at $0_{\mathbb{G}_2(\mathbb{R}^d)}$, can be extended in a usual manner to a left invariant $(0, 2)$-tensor, still denoted g, and defined on the whole Lie group $\mathbb{G}_2(\mathbb{R}^d)$. Precisely, for $x \in \mathbb{G}_2(\mathbb{R}^d)$, we define g_x on the tangent space to $\mathbb{G}_2(\mathbb{R}^d)$ at x in such a way that for $1 \leq i, j, k, l \leq d$,

$$g_x(D_i(x), D_j(x)) = 0 \text{ if } i \neq j,$$

$$g_x(D_i(x), D_i(x)) = 1,$$

$$g_x(D_i(x), [D_j, D_k](x)) = 0,$$

$$g_x([D_i, D_j](x), [D_k, D_l](x)) = 0.$$

Since g is not definite positive, the associated geometry is not Riemannian but sub-Riemannian. The relevant object for studying this geometry is the so-called horizontal distribution $\mathcal{H}$ which is defined as the smoothly varying family of vector spaces

$$\mathcal{H}_x = \mathbf{span} \left(D_1(x), ..., D_d(x) \right), \ x \in \mathbb{G}_2(\mathbb{R}^d).$$

An absolutely continuous curve $c : [0,1] \to \mathbb{G}_2(\mathbb{R}^d)$ is called horizontal if for almost every $s \in [0,1]$ we have $c'(s) \in \mathcal{H}_{c(s)}$.

Proposition 2.1   *Let $c = (\alpha, \omega)$ be an absolutely continuous curve $[0,1] \to \mathbb{G}_2(\mathbb{R}^d)$. It is a horizontal curve if and only if*

$$\omega'(s) = \frac{1}{2}\alpha(s) \wedge \alpha'(s).$$

Proof.   The curve c is horizontal if and only if there exist $\lambda_1, ..., \lambda_d$ such that

$$c'(s) = \sum_{i=1}^{d} \lambda_i(s) D_i(c(s)).$$

Since

$$D_i(x) = \frac{\partial}{\partial x^i} + \frac{1}{2}\sum_{j<i} x^j \frac{\partial}{\partial x^{j,i}} - \frac{1}{2}\sum_{j>i} x^j \frac{\partial}{\partial x^{i,j}}, \ 1 \le i \le d,$$

we obtain, first, by an identification of the coefficients in front of the $\frac{\partial}{\partial x^i}$'s

$$\lambda_i(s) = \alpha_i'(s).$$

The identification of the terms in front of the $\frac{\partial}{\partial x^{i,j}}$'s leads exactly to

$$\omega'(s) = \frac{1}{2}\alpha(s) \wedge \alpha'(s).$$

$\square$

The length of an horizontal curve c with respect to g is defined by

$$l(c) = \int_0^1 \sqrt{g_{c(s)}(c'(s), c'(s))} ds.$$

We can now turn to the first version of Chow theorem, whose proof shall be given later (see Theorem 2.4).

Theorem 2.2   *Given two points x and $y \in \mathbb{G}_2(\mathbb{R}^d)$, there is at least one horizontal curve $c : [0,1] \to \mathbb{G}_2(\mathbb{R}^d)$ such that $c(0) = x$ and $c(1) = y$.*

The Carnot-Carathéodory distance between x and y and denoted $d_g(x,y)$ is defined as being the infimum of the lengths of all the horizontal curves joining x and y. It is easily checked that this distance satisfies $d_g(c \cdot x, c \cdot y) = c d_g(x,y)$, for any $c > 0, x, y \in \mathbb{G}_2(\mathbb{R}^d)$, where the multiplication by c in the group corresponds to the scaling defined by the formula (2.2). A horizontal curve with length $d_g(x,y)$ is called a sub-Riemannian geodesic joining x and y.

Proposition 2.2   *Let $c = (\alpha, \omega)$ be a smooth horizontal curve $[0,1] \to$ $\mathbb{G}_2(\mathbb{R}^d)$. Then, it is a sub-Riemannian geodesic if and only if there exists a skew-symmetric $d \times d$ matrix Λ such that*

$$\alpha'' = \Lambda \alpha'.$$

Proof.   First, we note that the equation given in the proposition is nothing else than the Euler-Lagrange equation

$$\frac{d}{ds}\frac{\partial L}{\partial c'} = \frac{\partial L}{\partial c},$$

associated with the Lagrangian with constraints

$$L(c(s), c'(s)) = \frac{1}{2}g_{c(s)}(c'(s), c'(s)) + 2\sum_{i<j} \Lambda_{i,j}\theta^{i,j}(c'(s)),$$

where Λ is an arbitrary skew-symmetric $d \times d$ matrix and $\theta^{i,j} = dx^{i,j}$ is the one-form on $\mathfrak{g}_2(\mathbb{R}^d)$ which vanishes on $\mathfrak{g}_2(\mathbb{R}^d)$ excepted on the vector space spanned by $[D_i, D_j]$. Indeed, since c is horizontal we can write

$$c'(s) = \sum_{i=1}^{d} \alpha_i'(s)D_i(c(s)),$$

thus

$$L(c(s), c'(s)) = \frac{1}{2}\sum_{i=1}^{d}(\alpha_i'(s))^2 + \sum_{i<j} \Lambda_{i,j}\left(\alpha_i(s)\alpha_j'(s) - \alpha_j(s)\alpha_i'(s)\right),$$

and

$$\frac{\partial L}{\partial \alpha_i'} = \alpha_i'(s) - \sum_{j=1}^{d}\Lambda_{i,j}\alpha_j(s).$$

Now, to conclude the proof we have to check that all the sub-Riemannian geodesics are indeed critical points of the constrained Lagrangian that has been considered. Since the proof of this fact is not obvious and quite technical, we refer the interested reader to [Golé and Karidi (1995)]. $\square$

Remark 2.1   *It can be shown that in $\mathbb{G}_2(\mathbb{R}^d)$, all the sub-Riemannian geodesics are smooth (see e.g. [Golé and Karidi (1995)]).*

From the previous proposition, we deduce that the sub-Riemannian geodesics in $\mathbb{G}_2(\mathbb{R}^d)$ are *generalized helices*, i.e. curves with constant curvature. Precisely, let $c = (\alpha, \omega)$ be a sub-Riemannian geodesic. The curve c is smooth and solves the equation

$$\alpha'' = \Lambda\alpha', \tag{2.3}$$

for some skew-symmetric $d \times d$ matrix Λ. It is now an elementary fact in linear algebra that we can then write a decomposition

$$\mathbb{R}^d = \mathbf{Ker}(\Lambda) \oplus H_1 \oplus \cdots H_m$$

where $H_1,...,H_m$ are planes (i.e. two-dimensional spaces) such that $\Lambda H_i = H_i$. Therefore, a direct quadrature of the equation (2.3) shows that the projection of c onto any one of these planes is part of a circle, and its projection on the kernel is a line segment.

Now, since the geodesics of the geometry of $\mathbb{G}_2(\mathbb{R}^d)$ are known, it is natural to be interested in the shape of balls. We denote for $\varepsilon > 0$, $\mathbf{B}_g(0, \varepsilon)$ the open ball with radius ε for the Carnot-Carathéodory metric d_g. By direct but tedious computations, it can be shown that the sphere $\partial\mathbf{B}_g(0, \varepsilon)$ is not a smooth sub-manifold of $\mathbb{G}_2(\mathbb{R}^d)$. Precisely, $\partial\mathbf{B}_g(0, \varepsilon)$ has a singularity in each of the vertical directions $[D_i, D_j] = \frac{\partial}{\partial x^{i,j}}$. For instance in the Heisenberg group, i.e. in $\mathbb{G}_2(\mathbb{R}^2)$, the unit sphere looks like an apple (see Fig. 2.1) whose parametric equations are

$$x(u, v) = \frac{\cos u \sin v - \sin u(1 - \cos v)}{v},$$
$$y(u, v) = \frac{\sin u \sin v + \cos u(1 - \cos v)}{v},$$
$$z(u, v) = \frac{v - \sin v}{v^2},$$

where $u \in [0, 2\pi]$, $v \in [-2\pi, 2\pi]$. By letting the parameter v describe $\mathbb{R}$, we obtain the wave front of the Heisenberg group, whose structure is quite complicated (see Fig. 2.2 which represents half of the wave front).

For the Euclidean metric of $\mathbb{G}_2(\mathbb{R}^d) = \mathbb{R}^d \times \mathbb{R}^{\frac{d(d-1)}{2}}$, the size of the sphere $\mathbf{B}_g(0, \varepsilon)$ in the horizontal directions is approximatively ε and approximatively ε^2 in the vertical directions. Precisely, as it will be seen later, if for $\varepsilon > 0$, we denote

$$\mathbf{Box}(\varepsilon) = \{(\alpha, \omega) \in \mathbb{G}_2(\mathbb{R}^d), |\,\alpha_i\,| \le \varepsilon, |\,\omega_{j,k}\,| \le \varepsilon^2\},$$

there exist positive constants c_1 and c_2 and ε_0 such that for any $0 < \varepsilon < \varepsilon_0$,

$$\mathbf{Box}(c_1\varepsilon) \subset \mathbf{B}_g(0, \varepsilon) \subset \mathbf{Box}(c_2\varepsilon).$$

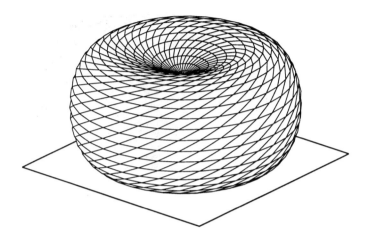

Fig. 2.1 The unit Heisenberg sphere.

From this estimation, we deduce immediately that the topology given by the distance d_g is compatible with the natural topology of the Lie group $\mathbb{G}_2(\mathbb{R}^d)$. We can also deduce that the Hausdorff dimension of the metric space $(\mathbb{G}_2(\mathbb{R}^d), d_g)$ is $\dim \mathcal{V}_1 + 2 \dim \mathcal{V}_2 = d^2$ which is quite striking since the dimension of $\mathbb{G}_2(\mathbb{R}^d)$ as a topological manifold is only equal to $\frac{d(d+1)}{2}$. For further details on these properties of the metric space $(\mathbb{G}_2(\mathbb{R}^d), d_g)$, we refer to the next section.

We now come back to the study of the stochastic differential equation

$$X_t^{x_0} = x_0 + \sum_{i=1}^{d} \int_0^t V_i(X_s^{x_0}) \circ dB_s^i, \ t \geq 0, \qquad (2.4)$$

where $x_0 \in \mathbb{R}^n$, $V_1, ..., V_d$ are C^∞ bounded vector fields on $\mathbb{R}^n$ and $(B_t)_{t \geq 0}$ is a d-dimensional standard Brownian motion and show the *universality* of the group $\mathbb{G}_2(\mathbb{R}^d)$ in the study of two-step nilpotent diffusions.

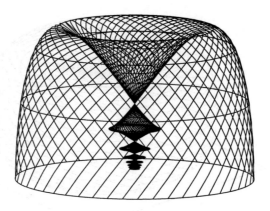

Fig. 2.2   The wave front in the Heisenberg group.

Theorem 2.3   *Assume that the Lie algebra $\mathfrak{L} = \mathbf{Lie}(V_1, ..., V_d)$ is two-step nilpotent. There exists a smooth map*

$$F : \mathbb{R}^n \times \mathbb{G}_2(\mathbb{R}^d) \to \mathbb{R}^n$$

such that, for $x_0 \in \mathbb{R}^n$, the solution $(X_t^{x_0})_{t \geq 0}$ of the stochastic differential equation (2.4) can be written

$$X_t^{x_0} = F(x_0, B_t^*),$$

where $(B_t^*)_{t \geq 0}$ is the lift of $(B_t)_{t \geq 0}$ in the group $\mathbb{G}_2(\mathbb{R}^d)$.

Proof.   Let us consider the function $F : \mathbb{R}^n \times \mathbb{G}_2(\mathbb{R}^d)$ defined for $x \in \mathbb{R}^n$ and $g = (\alpha, \omega) \in \mathbb{G}_2(\mathbb{R}^d)$ by

$$F(x, g) = \exp \left(\sum_{i=1}^{d} \alpha^i V_i + \frac{1}{2} \sum_{1 \leq i < j \leq d} [V_i, V_j] \omega^{i,j} \right) (x).$$

Since the Stratonovitch integration satisfies the usual change of variable formula, an iteration of Itô's formula shows, by using Chen's development theorem, that the process $(F(x_0, B_t^*))_{t \geq 0}$ solves the equation (2.4). We conclude by the pathwise uniqueness property. $\quad\square$

The conclusion of this section is hence the following: The natural geometry associated with a 2-step nilpotent stochastic differential equation is the sub-Riemannian geometry of a quotient of the free 2-step Carnot group $\mathbb{G}_2(\mathbb{R}^d)$. The next step for us is now to generalize the previous study to any nilpotent stochastic differential equation.

2.3 N-step nilpotent SDE's

We introduce now the notion of Carnot group. Carnot groups are to sub-Riemannian geometry what Euclidean spaces are to Riemannian geometry. Numerous papers and several books are devoted to the analysis of these groups (see e.g. [Bellaïche (1996)], [Folland and Stein (1982)], [Goodman (1976)], [Gromov (1996)]).

Definition 2.3   A Carnot group of step (or depth) N is a simply connected Lie group $\mathbb{G}$ whose Lie algebra can be written

$$\mathcal{V}_1 \oplus ... \oplus \mathcal{V}_N,$$

where

$$[\mathcal{V}_i, \mathcal{V}_j] = \mathcal{V}_{i+j}$$

and

$$\mathcal{V}_s = 0, \text{ for } s > N.$$

Example 2.2   The group $(\mathbb{R}^d, +)$ is the only commutative Carnot group.

Example 2.3   The group $\mathbb{G}_2(\mathbb{R}^d)$ considered in the previous section is a Carnot group of depth 2.

Example 2.4   Consider the set $\mathbb{H}_n = \mathbb{R}^{2n} \times \mathbb{R}$ endowed with the group law

$$(x, \alpha) \star (y, \beta) = \left(x + y, \alpha + \beta + \frac{1}{2}\omega(x, y) \right),$$

where ω is the standard symplectic form on $\mathbb{R}^{2n}$, that is

$$\omega(x, y) = x^t \begin{pmatrix} 0 & -\mathbf{I}_n \\ \mathbf{I}_n & 0 \end{pmatrix} y.$$

Observe that $\mathbb{H}_1$ is the Heisenberg group. On $\mathfrak{h}_n$ the Lie bracket is given by

$$[(x, \alpha), (y, \beta)] = (0, \omega(x, y)),$$

and it is easily seen that

$$\mathfrak{h}_n = \mathcal{V}_1 \oplus \mathcal{V}_2,$$

where $\mathcal{V}_1 = \mathbb{R}^{2n} \times \{0\}$ and $\mathcal{V}_2 = \{0\} \times \mathbb{R}$. Therefore $\mathbb{H}_n$ is a Carnot group of depth 2.

We consider throughout this section a Lie group $\mathbb{G}$ which satisfies the hypothesis of the above definition. Notice that the vector space $\mathcal{V}_1$, which is called the basis of $\mathbb{G}$, Lie generates $\mathfrak{g}$, where $\mathfrak{g}$ denotes the Lie algebra of $\mathbb{G}$. Since $\mathbb{G}$ is step N nilpotent and simply connected, the exponential map is a diffeomorphism and the Baker-Campbell-Hausdorff formula therefore completely characterizes the group law of $\mathbb{G}$ because for $U, V \in \mathfrak{g}$,

$$\exp U \exp V = \exp \left(P(U, V) \right)$$

for some universal Lie polynomial P whose first terms are given by

$$P(U, V) = U + V + \tfrac{1}{2}[U, V] + \tfrac{1}{12}[[U, V], V] - \tfrac{1}{12}[[U, V], U] \\ - \tfrac{1}{48}[V, [U, [U, V]]] - \tfrac{1}{48}[U, [V, [U, V]]] + \cdots.$$

(see Appendix B for an explicit formula). On $\mathfrak{g}$ we can consider the family of linear operators $\delta_t : \mathfrak{g} \to \mathfrak{g}$, $t \geq 0$ which act by scalar multiplication t^i on $\mathcal{V}_i$. These operators are Lie algebra automorphisms due to the grading. The maps δ_t induce Lie group automorphisms $\Delta_t : \mathbb{G} \to \mathbb{G}$ which are called the canonical dilations of $\mathbb{G}$. Let us now take a basis $U_1, ..., U_d$ of the vector space $\mathcal{V}_1$. The vectors U_i's can be seen as left invariant vector fields on $\mathbb{G}$ so that we can consider the following stochastic differential equation on $\mathbb{G}$:

$$d\tilde{B}_t = \sum_{i=1}^{d} \int_0^t U_i(\tilde{B}_s) \circ dB_s^i, \ t \geq 0, \tag{2.5}$$

which is easily seen to have a unique (strong) solution $(\tilde{B}_t)_{t \geq 0}$ associated with the initial condition $\tilde{B}_0 = 0_{\mathbb{G}}$.

Definition 2.4   The process $(\tilde{B}_t)_{t \geq 0}$ is called the lift of the Brownian motion $(B_t)_{t \geq 0}$ in the group $\mathbb{G}$ with respect to the basis $(U_1, ..., U_d)$.

Remark 2.2 *Notice that $(\tilde{B}_t)_{t \geq 0}$ is a Markov process with generator $\frac{1}{2} \sum_{i=1}^{d} U_i^2$. This second-order differential operator is, by construction, left-invariant and hypoelliptic.*

Proposition 2.3 *We have*

$$\tilde{B}_t = \exp \left(\sum_{k=1}^{N} \sum_{I=(i_1,\dots,i_k)} \Lambda_I(B)_t U_I \right), \ t \geq 0,$$

where:

$$\Lambda_I(B)_t = \sum_{\sigma \in \mathfrak{S}_k} \frac{(-1)^{e(\sigma)}}{k^2 \binom{k-1}{e(\sigma)}} \int_{\Delta^k[0,t]} \circ dB^{\sigma^{-1} \cdot I}.$$

Proof. This is obviously a straightforward consequence of the Chen-Strichartz development Theorem 1.1, whose proof is given in Chapter 1. Nevertheless, since the general proof of Theorem 1.1 is purely algebraic we believe that it is interesting to hint a more *pathwise oriented* proof. We follow closely [Strichartz (1987)] and proceed in two steps. In the first step we show that if $\omega : \mathbb{R}_{\geq 0} \to \mathbb{R}^d$ is an absolutely continuous path, then the solution of the ordinary differential equation

$$dx_t = \sum_{i=1}^{d} U_i(x_t) d\omega_t^i$$

is given by

$$x_t = \exp \left(\sum_{k=1}^{N} \sum_{I=\{i_1,\dots,i_k\}} \Lambda_I(\omega)_t U_I \right),$$

where

$$\Lambda_I(\omega)_t = \sum_{\sigma \in \mathfrak{S}_k} \frac{(-1)^{e(\sigma)}}{k^2 \binom{k-1}{e(\sigma)}} \int_{\Delta^k[0,t]} d\omega^{\sigma^{-1} \cdot I}.$$

In the second step, we observe that Stratonovitch differentiation follows the same rules as the differentiation of absolutely continuous paths.

Step 1. Let $\omega : \mathbb{R}_{\geq 0} \to \mathbb{R}^d$ be an absolutely continuous path. Let $t > 0$ and $n \in \mathbb{N}^*$. We consider a regular subdivision $0 = t_0 < t_1 < \dots < t_n = t$

of the time interval $[0, t]$. The linear continuous interpolation of the path $(\omega_s)_{0 \le s \le t}$ is given by:

$$\tilde{\omega}_s^k = n(\omega_{t_{i+1}}^k - \omega_{t_i}^k)(s - t_i) + \omega_{t_i}^k,$$

where $s \in [t_i, t_{i+1})$, $k = 1, ..., d$. If $(\tilde{x}_s)_{0 \le s \le t}$ denotes the solution of the ordinary differential equation

$$d\tilde{x}_s = \sum_{k=1}^{d} U_k(\tilde{x}_s) d\tilde{\omega}_s^k, \ 0 \le s \le t,$$

it is easily seen that

$$\tilde{x}_t = \exp\left(\sum_{k=1}^{d}(\omega_t^k - \omega_{t_{n-1}}^k)U_k\right) \cdots \exp\left(\sum_{k=1}^{d}(\omega_{t_1}^k - \omega_{t_0}^k)U_k\right).$$

Now we use the Baker-Campbell-Hausdorff formula (see Appendix B) to write the previous product of exponentials under the form

$$\tilde{x}_t = \exp\left(\sum_{k=1}^{N} \sum_{I=\{i_1, ..., i_k\}} \Lambda_I^n(\omega)_t U_I\right),$$

and similar arguments, using furthermore the convergence of Riemann sums, to those given in the proof of the Chen-Strichartz formula show that

$$\Lambda_I^n(\omega)_t \to_{n \to +\infty} \Lambda_I(\omega)_t.$$

Since

$$\tilde{x}_t \to_{n \to +\infty} x_t,$$

we conclude that

$$x_t = \exp\left(\sum_{k=1}^{N} \sum_{I=\{i_1, ..., i_k\}} \Lambda_I(\omega)_t U_I\right).$$

Step 2. Since the Itô's formula with Stratonovitch integrals has the same form as the usual change of variable formula, an iteration of Itô's formula shows the expected result. $\square$

Observe that, due to the elementary fact

$$(\Lambda_I(B)_{ct})_{t \ge 0} =^{law} \left(c^{\frac{|I|}{2}} \Lambda_I(B)_t\right)_{t \ge 0},$$

we have the following scaling property, for every $c > 0$

$$\left(\tilde{B}_{ct}\right)_{t \geq 0} \overset{\text{law}}{=} \left(\Delta_{\sqrt{c}}\tilde{B}_t\right)_{t \geq 0}.$$

This scaling property leads directly to the following value at $0_{\mathbb{G}}$ of the density $\tilde{p}_t$ of $\tilde{B}_t$ with respect to any Haar measure of $\mathbb{G}$:

$$\tilde{p}_t (0_{\mathbb{G}}) = \frac{C}{t^{\frac{D}{2}}}, \ t > 0,$$

where $C > 0$ and $D = \sum_{i=1}^{N} i \dim \mathcal{V}_i$. Moreover, from [Alexopoulos and Lohoué (2004)], for $\tilde{p}_t$, $t > 0$, the following estimates hold:

Proposition 2.4 *For $(i_1, ..., i_p) \in \{1, ..., d\}^p$, $g \in \mathbb{G}$, $q \geq 0$, $t > 0$,*

$$\mid U_{i_1} \cdots U_{i_p} \partial_t^q \tilde{p}_t(g) \mid \leq \frac{A_{p,q}}{t^{\frac{D+p+2q}{2}}} \exp\left(-\frac{d_g(0_{\mathbb{G}}, g)^2}{B_{p,q}t}\right),$$

where $A_{p,q}$ and $B_{p,q}$ are non-negative constants.

We now turn to the geometry of $\mathbb{G}$. The Lie algebra $\mathfrak{g}$ can be identified with the set of left-invariant vector fields on $\mathbb{G}$. From this identification and from the decomposition

$$\mathfrak{g} = \mathcal{V}_1 \oplus ... \oplus \mathcal{V}_N,$$

we deduce a decomposition of the tangent space $T_x\mathbb{G}$ to $\mathbb{G}$ at $x \in \mathbb{G}$:

$$T_x\mathbb{G} = \mathcal{V}_1(x) \oplus ... \oplus \mathcal{V}_N(x),$$

where $\mathcal{V}_i(x)$ is the fiber at x of the left-invariant differential system spanned by $\mathcal{V}_i$. This decomposition endows naturally $\mathbb{G}$ with a left-invariant $(0, 2)$-tensor g. Precisely, for $x \in \mathbb{G}$, we define g_x as being the scalar product on $T_x\mathbb{G}$ such that:

(1) The vectors $U_1(x), ..., U_d(x)$ form an orthonormal basis;
(2) $g_x \mid_{\mathcal{V}_i(x) \times \mathcal{V}_j(x)} = 0$, if i or j is different from 1.

An absolutely continuous curve $c : [0, 1] \to \mathbb{G}$ is called horizontal if for almost every $s \in [0, 1]$ we have $c'(s) \in \mathcal{V}_1(c(s))$. The length of a horizontal curve c with respect to g is defined by

$$l(c) = \int_0^1 \sqrt{g_{c(s)}(c'(s), c'(s))}ds.$$

We can now state the basic result on the geometry of Carnot groups: The Chow's theorem.

Theorem 2.4   *Given two points x and $y \in \mathbb{G}$, there is at least one horizontal absolutely continuous curve $c : [0,1] \to \mathbb{G}$ such that $c(0) = x$ and $c(1) = y$.*

Proof.   Let us denote G the subgroup of diffeomorphisms $\mathbb{G} \to \mathbb{G}$ generated by the one-parameter subgroups corresponding to $U_1, ..., U_d$. The Lie algebra of G can be identified with the Lie algebra generated by $U_1, ..., U_d$, i.e. $\mathfrak{g}$. We deduce that G can be identified with $\mathbb{G}$ itself, so that it acts transitively on $\mathbb{G}$. It means that for every $x \in \mathbb{G}$, the map $G \to \mathbb{G}$, $g \to g(x)$ is surjective. Thus, every two points in $\mathbb{G}$ can be joined by a piecewise smooth horizontal curve where each piece is a segment of an integral curve of one of the vector fields U_i. $\square$

Remark 2.3   *In the above proof, the horizontal curve constructed to join two points is not smooth. Nevertheless, it can be shown that it is always possible to connect two points with a smooth horizontal curve (see [Gromov (1996)] pp. 120).*

The Carnot-Carathéodory distance between x and y and denoted $d_g(x,y)$ is defined as being the infimum of the lengths of all the horizontal curves joining x and y. It is easily checked that this distance satisfies $d_g(\Delta_c x, \Delta_c y) = c d_g(x,y)$, for every $c > 0, x, y \in \mathbb{G}$.

Remark 2.4   *The distance d_g depends on the choice of a basis for $\mathcal{V}_1$. Nevertheless, all the Carnot-Carathéodory distances that can be constructed are bi-Lipschitz equivalent.*

A horizontal curve with length $d_g(x,y)$ is called a sub-Riemannian geodesic joining x and y. The topology of the metric space $(\mathbb{G}, d_g)$ is really of interest. The basic tool to investigate this topology is the so-called ball-box theorem. Denote for $\varepsilon > 0$,

$$\mathbf{Box}(\varepsilon) = \{ g \in \mathfrak{g}, \| g_i \|_i \leq \varepsilon^i, 1 \leq i \leq N \},$$

where g_i denotes the projection of g on $\mathcal{V}_i$ and $\| g_i \|_i$ the norm of g_i with respect to any norm of $\mathcal{V}_i$ (all the norms on $\mathcal{V}_i$ are equivalent). We have then the following so-called ball-box theorem.

Proposition 2.5   *There exist positive constants c_1 and c_2 and ε_0 such that for any $0 < \varepsilon < \varepsilon_0$,*

$$\exp\left(\mathbf{Box}(c_1 \varepsilon)\right) \subset \mathbf{B}_g(0, \varepsilon) \subset \exp\left(\mathbf{Box}(c_2 \varepsilon)\right),$$

where $\mathbf{B}_g(0, \varepsilon)$ *denotes the open ball with radius* ε *in the metric space* $(\mathbb{G}, d_g)$.

Remark 2.5   *Let us mention that the previous proposition is actually a consequence of the equivalence of all the homogeneous norms on* $\mathbb{G}$. *A homogeneous norm on* $\mathbb{G}$ *is a continuous function* $\| \cdot \| \colon \mathbb{G} \to [0, +\infty)$, *smooth away from the origin, such that:*

(1) $\| \Delta_c x \| = c \| x \|$, $c > 0$, $x \in \mathbb{G}$;
(2) $\| x^{-1} \| = \| x \|$, $x \in \mathbb{G}$;
(3) $\| x \| = 0$ *if and only if* $x = 0_{\mathbb{G}}$.

For analytical questions in Carnot groups (estimations,...) it is often more convenient to work with a given homogeneous norm rather than with the Carnot-Carathéodory distance itself.

We have important consequences of the ball-box theorem. First, the topology given by the distance d_g is compatible with the natural topology of the Lie group $\mathbb{G}$. Secondly, we can compute the Hausdorff dimension of the metric space $(\mathbb{G}, d_g)$.

To make what follows clear, let us recall some facts about Hausdorff measure and Hausdorff dimension (see e.g. [Falconer (1986)] for a detailed account of this material). Let $(\mathbb{M}, d)$ be a metric space, and let Ω be an open subset of $\mathbb{M}$. Consider an open cover $\mathcal{U} = \{U_\alpha\}$ of Ω and set for $s > 0$

$$\mu^s(\Omega, \mathcal{U}) = \sum_\alpha (\mathrm{diam}(U_\alpha))^s \,.$$

This quantity (possibly infinite) is called the approximate s-dimensional Hausdorff measure. We will use the notation $| \mathcal{U} | < \epsilon$ to mean that each U_α has diameter less than ϵ. The ϵ-approximate s-dimensional measure of Ω is the number

$$\mu^s_\epsilon(\Omega) = \inf\{\mu^s(\Omega, \mathcal{U}), | \mathcal{U} | < \epsilon\}.$$

Finally, set

$$\mu^s(\Omega) = \lim_{\epsilon \to 0} \mu^s_\epsilon(\Omega).$$

This quantity is called the s-dimensional Hausdorff measure of Ω.

Proposition 2.6   *There is a unique value D, called the Hausdorff dimension of the open set Ω, with the property that $\mu^s(\Omega) = +\infty$ for all $s < D$ and $\mu^s(\Omega) = 0$ for all $s > D$.*

The metric space $(\mathbb{M}, d)$ is said to have an Hausdorff dimension equal to D, if any open set of $\mathbb{M}$ has an Hausdorff dimension equal to D.

Proposition 2.7   *The Hausdorff dimension of the metric space $(\mathbb{G}, d_g)$ is equal to*

$$D = \sum_{j=1}^{N} j \dim \mathcal{V}_j.$$

Proof.   We shall show that the unit ball $\mathbf{B}_g(0,1)$ has Hausdorff dimension $D = \sum_{j=1}^{N} j \dim \mathcal{V}_j$. We know that, due to the dilations, the volume of a ball $\mathbf{B}_g(0,\varepsilon)$ is $C\varepsilon^D$. Consider now a maximal filling of $\mathbf{B}_g(0,1)$ with balls of radius ε. An upper bound for the number of balls $N(\varepsilon)$ in this filling is

$$N(\varepsilon) \leq \frac{1}{\varepsilon^D}.$$

Now, the set of concentric balls of radius 2ε constructed from this filling covers $\mathbf{B}_g(0,1)$. Each of these balls has diameter smaller than 4ε, thus the Hausdorff s-dimensional measure of $\mathbf{B}_g(0,1)$ is smaller than

$$\lim_{\varepsilon \to 0} N(\varepsilon)^s,$$

which is 0 if $s > D$. Therefore the Hausdorff dimension is smaller than D.

Conversely, given any covering of $\mathbf{B}_g(0,1)$ by sets of diameter $\leq \varepsilon$, there is an associated covering with balls of the same diameter. The number $M(\varepsilon)$ of these balls has the lower bound:

$$M(\varepsilon) \geq \frac{1}{\varepsilon^D}.$$

We deduce that for every $s > 0$,

$$\sum_{cover} \varepsilon^s \geq \frac{\varepsilon^s}{\varepsilon^D} = \varepsilon^{s-D},$$

which shows that if $s < D$ then the Hausdorff s-dimensional measure of $\mathbf{B}_g(0,1)$ is $+\infty$. Therefore the Hausdorff dimension is greater than D $\square$

Remark 2.6   *Observe, and this is typical in sub-Riemannian geometry, that the Hausdorff dimension of $\mathbb{G}$ is therefore strictly greater than the topological dimension.*

Carnot groups have their own concept of differentiation. In order to present this concept introduced in [Pansu (1989)], we first have to define a concept of linear maps between two Carnot groups.

Let $\mathbb{G}_1$ and $\mathbb{G}_2$ be two Carnot groups with Lie algebras $\mathfrak{g}_1$ and $\mathfrak{g}_2$. A Lie group morphism $\phi : \mathbb{G}_1 \to \mathbb{G}_2$ is said to be a Carnot group morphism if for any $t \geq 0$, $g \in \mathbb{G}_1$,

$$\phi(\Delta_t^{\mathbb{G}_1} g) = \Delta_t^{\mathbb{G}_2} \phi(g),$$

where $\Delta^{\mathbb{G}_1}$ (resp. $\Delta^{\mathbb{G}_2}$) denote the canonical dilations on $\mathbb{G}_1$ (resp. $\mathbb{G}_2$). In the same way, a Lie algebra morphism $\alpha : \mathfrak{g}_1 \to \mathfrak{g}_2$ is said to be a Carnot algebra morphism if for any $t \geq 0$, $x \in \mathfrak{g}_1$,

$$\alpha(\delta_t^{\mathfrak{g}_1} x) = \delta_t^{\mathfrak{g}_2} \alpha(x),$$

where $\delta^{\mathbb{G}_1}$ (resp. $\delta^{\mathbb{G}_2}$) denote the canonical dilations on $\mathfrak{g}_1$ (resp. $\mathfrak{g}_2$). Observe that if ϕ is a Carnot group morphism, then the derivative $d\phi$ is a Carnot algebra morphism.

Let now $F : \mathbb{G}_1 \to \mathbb{G}_2$ be a map. When it exists, the Pansu's derivative $d_P F$, is defined by

$$d_P F(g)(h) = \lim_{t \to 0} \left(\Delta_t^{\mathbb{G}_2} \right)^{-1} \left(F(g)^{-1} F \left(g \Delta_t^{\mathbb{G}_1} h \right) \right), \quad g, h \in \mathbb{G}_1.$$

and $d_P F(g)$ is a Carnot group morphism.

In this setting, Pansu has proved a deep generalization of the classical Rademacher's theorem which asserts that a Lipschitz map between Euclidean spaces is almost everywhere differentiable. Indeed, any Lipschitz map (with respect to the Carnot-Carathéodory distances of $\mathbb{G}_1$ and $\mathbb{G}_2$) is almost everywhere Pansu differentiable. An important consequence of this is that can be no bi-Lipschitz map between Carnot groups that are not isomorphic as groups.

We conclude now our presentation of the Carnot groups with the free Carnot groups. The Carnot group $\mathbb{G}$ is said to be free if $\mathfrak{g}$ is isomorphic to the nilpotent free Lie algebra with d generators. In that case, dim $\mathcal{V}_j$ is the number of Hall words of length j in the free algebra with d generators (see Appendix B). We thus have, according to [Bourbaki (1972)] or [Reutenauer (1993)] pp.96:

$$\dim \mathcal{V}_j = \frac{1}{j} \sum_{i|j} \mu(i) d^{\frac{j}{i}}, \ j \leq N,$$

where μ is the Möbius function. We easily deduce from this that when $N \to +\infty$,

$$\dim \mathfrak{g} \sim \frac{d^N}{N}.$$

An important algebraic point is that there are many algebraically non iso-morphic Carnot groups having the same dimension (even uncountably many for $n \geq 6$), but up to an isomorphism there is one and only one free Carnot with a given depth and a given dimension for the basis. Actually, as in the theory of vector spaces, we can reduce the study of the free Carnot groups to standard numerical models. Let us denote $m = \dim \mathbb{G}$. Choose now a Hall family and consider the $\mathbb{R}^m$-valued semimartingale $(B_t^*)_{t \geq 0}$ obtained by writing the components of $(\ln(\tilde{B}_t))_{t \geq 0}$ in the corresponding Hall basis of $\mathfrak{g}$. It is easily seen that $(B_t^*)_{t \geq 0}$ solves a stochastic differential equation that can be written

$$B_t^* = \sum_{i=1}^d \int_0^t D_i(B_s^*) \circ dB_s^i,$$

where the D_i's are polynomial vector fields on $\mathbb{R}^m$ (for an explicit form of the D_i's, which depend of the choice of the Hall basis, we refer to [Ger-shkovich and Vershik (1994)] pp.27) . With these notations, we have the following proposition which results of our very construction.

Proposition 2.8   *On $\mathbb{R}^m$, there exists a unique group law $\circledast$ which makes the vector fields $D_1, ..., D_d$ left invariant. This group law is, unimodular[1], polynomial of degree N and we have moreover*

$$(\mathbb{R}^m, \circledast) \sim \mathbb{G}.$$

The group $(\mathbb{R}^m, \circledast)$ is called the free Carnot group of step N over $\mathbb{R}^d$. It shall be denoted $\mathbb{G}_N(\mathbb{R}^d)$ and its Lie algebra $\mathfrak{g}_N(\mathbb{R}^d)$. The process $(B_t^*)_{t \geq 0}$ shall be called the lift of $(B_t)_{t \geq 0}$ in $\mathbb{G}_N(\mathbb{R}^d)$ with respect to the basis $(D_1, ..., D_d)$.

Remark 2.7   *The important point with $\mathfrak{g}_N(\mathbb{R}^d)$ is that this is not an abstract Lie algebra. This is a Lie algebra of vector fields defined on $\mathbb{R}^m$.*

Remark 2.8   *By construction of $(\mathbb{G}_N(\mathbb{R}^d), \circledast)$ the exponential map is simply the identity.*

The universality of $\mathbb{G}_N(\mathbb{R}^d)$ is the following.

Proposition 2.9   *Let $\mathbb{G}$ be any Carnot group. There exists a Carnot group surjective morphism $\pi : \mathbb{G}_N(\mathbb{R}^d) \to \mathbb{G}$, where d is the dimension of the basis of $\mathbb{G}$ and N its depth.*

---

[1]A group law on $\mathbb{R}^m$ is said to be unimodular if the translations let the Lebesgue measure invariant.

Proof. Let $U_1, ..., U_d$ be a basis of the basis of $\mathbb{G}$. Since $\mathbb{G}_N(\mathbb{R}^d)$ is free, there exists a unique Lie algebra surjective morphism $d\pi : \mathfrak{g}_N(\mathbb{R}^d) \to \mathfrak{g}$ such that $d\pi(D_i) = U_i$, $i = 1, ..., d$. We can now define a surjective Carnot group morphism $\pi : \mathbb{G}_N(\mathbb{R}^d) \to \mathbb{G}$ by $\pi(e^g) = e^{d\pi(g)}$, $g \in \mathfrak{g}_N(\mathbb{R}^d)$. Observe that it defines π in a unique way because in Carnot groups the exponential map is a diffeomorphism. $\qquad\square$

Notice that $\mathbb{G}_N(\mathbb{R}^d)$ is, by construction, endowed with the basis of vector fields $(D_1, ..., D_d)$. These vector fields agree at the origin with $\left(\frac{\partial}{\partial x_1}, \cdots, \frac{\partial}{\partial x_d} \right)$. To make our approach essentially frame independent, it is important to relate the horizontal lifts of the same Brownian motion with respect to two different basis.

Proposition 2.10   *Let $\varphi : \mathbb{R}^d \to \mathbb{R}^d$ be a vector space isomorphism. Let us denote $\hat{B}^*$ the horizontal lift of B in the group $\mathbb{G}_N(\mathbb{R}^d)$ with respect to the basis $(\varphi(D_1), ..., \varphi(D_d))$. Then there exists a unique Carnot group isomorphism $T_\varphi : \mathbb{G}_N(\mathbb{R}^d) \to \mathbb{G}_N(\mathbb{R}^d)$, such that for every $t \geq 0$:*

$$\hat{B}_t^* = T_\varphi(B_t^*).$$

Proof.   Since $\mathfrak{g}_N(\mathbb{R}^d)$ is free, φ can be extended in a unique way in a Lie algebra isomorphism $\mathfrak{g}_N(\mathbb{R}^d) \to \mathfrak{g}_N(\mathbb{R}^d)$. Notice that this extension commutes with the automorphisms δ_c, $c > 0$. Now, since $\mathbb{G}_N(\mathbb{R}^d)$ is simply connected and nilpotent, we can define a group automorphism $T_\varphi : \mathbb{G}_N(\mathbb{R}^d) \to \mathbb{G}_N(\mathbb{R}^d)$ by the property

$$T_\varphi(\exp x) = \exp(\varphi(x)), \; x \in \mathfrak{g}_N(\mathbb{R}^d).$$

It is then easily checked that for every $t \geq 0$:

$$\hat{B}_t^* = T_\varphi(B_t^*).$$

and moreover that T_φ commutes with the dilations Δ_c, $c > 0$. $\qquad\square$

The maps T_φ give the formulas for a change of basis in the theory of free Carnot groups. Notice that without the freeness assumption, T_φ may fail to exist.

Remark 2.9   *Notice that the map $\mathbf{Aut}(\mathbb{R}^d) \to \mathbf{Aut}(\mathbb{G}_N(\mathbb{R}^d))$, $\varphi \to T_\varphi$, is a group morphism.*

Finally, we now come back the study of the stochastic differential equation (2.1) and assume that the Lie algebra $\mathfrak{L} = \mathbf{Lie}(V_1, ..., V_d)$ is nilpotent of depth N, i.e. every commutator constructed from the V_i's with length greater than N is 0.

Theorem 2.5   *There exists a smooth map*

$$F : \mathbb{R}^n \times \mathbb{G}_N(\mathbb{R}^d) \to \mathbb{R}^n$$

such that, for $x_0 \in \mathbb{R}^n$, the solution $(X_t^{x_0})_{t \geq 0}$ of the stochastic differential equation (2.1) can be written

$$X_t^{x_0} = F(x_0, B_t^*),$$

where $(B_t^*)_{t \geq 0}$ is the lift of $(B_t)_{t \geq 0}$ in the group $\mathbb{G}_N(\mathbb{R}^d)$.

Proof.   This is a straightforward extension of Theorem 2.3. Indeed, as before, we notice that the Stratonovitch integration satisfies the usual change of variable formula, so that an iteration of Itô's formula shows, that the process

$$\left[\exp \left(\sum_{k=1}^{N} \sum_{I=(i_1,\ldots,i_k)} \Lambda_I(B)_t V_I \right) \right](x_0),$$

solves the equation (2.1). We deduce hence by pathwise uniqueness property that

$$X_t^{x_0} = \left[\exp \left(\sum_{k=1}^{N} \sum_{I=(i_1,\ldots,i_k)} \Lambda_I(B)_t V_I \right) \right](x_0).$$

The definition of $\mathbb{G}_N(\mathbb{R}^d)$ shows that we can therefore write

$$X_t^{x_0} = F(x_0, B_t^*).$$

$\square$

Remark 2.10   *Observe that this theorem implies the following inclusion of filtrations: for $t > 0$,*

$$\sigma(X_t^{x_0}) \subset \sigma\left(\Lambda_I(B)_t, \ |I| \leq N\right),$$

where $\sigma(X_t^{x_0})$ denotes the smallest σ-algebra containing $X_t^{x_0}$, and $\sigma\left(\Lambda_I(B)_t, \ |I| \leq N\right)$ denotes the smallest σ-algebra containing all the functionals $\Lambda_I(B)_t$ with I, word of length smaller than N.

The above theorem shows the universal property of $\mathbb{G}_N(\mathbb{R}^d)$ in theory of nilpotent stochastic flows. Let us mention its straightforward counterpart in the theory of second order hypoelliptic operators (this property is implicitly pointed out in the seminal work [Rotschild and Stein (1976)]).

Proposition 2.11 *Let*

$$\mathcal{L} = \sum_{i=1}^{d} V_i^2$$

be a second order differential operator on $\mathbb{R}^n$. Assume that the Lie algebra $\mathfrak{L} = \mathbf{Lie}\,(V_1, ..., V_d)$ which is generated by the vector fields V_i's admits a stratification

$$\mathcal{E}_1 \oplus \cdots \oplus \mathcal{E}_N,$$

with $\mathcal{E}_1 = \mathbf{span}\,(V_1, ..., V_d)$, $[\mathcal{E}_1, \mathcal{E}_i] = \mathcal{E}_{i+1}$ and $[\mathcal{E}_1, \mathcal{E}_N] = 0$. Then, there exists a submersion map $\pi : \mathbb{R}^m \to \mathbb{R}^n$, with $m = \dim \mathbb{G}_N(\mathbb{R}^d)$ such that for every smooth $f : \mathbb{R}^n \to \mathbb{R}$,

$$\Delta_{\mathbb{G}_N(\mathbb{R}^d)}\,(f \circ \pi) = (\mathcal{L}f) \circ \pi,$$

where $\Delta_{\mathbb{G}_N(\mathbb{R}^d)} = \sum_{i=1} D_i^2$, is the canonical sublaplacian on $\mathbb{G}_N(\mathbb{R}^d)$.

2.4 Pathwise approximation of solutions of SDEs

We now come back to the general case where the Lie algebra $\mathfrak{L} = \mathbf{Lie}\,(V_1, ..., V_d)$ is not nilpotent anymore. In that case it is, of course, not possible in full generality to represent $(X_t^{x_0})_{t \geq 0}$ in a deterministic way from the lift of the driving Brownian motion $(B_t)_{t \geq 0}$ in a finite dimensional Lie group. Nevertheless, the following result which is a consequence of a result due to [Castell (1993)] (see also [Ben Arous (1989b)]) shows that it is still possible to provide good pathwise approximations with lifts in the free Carnot groups.

Theorem 2.6 *Let $N \geq 1$. There exists a smooth map*

$$F : \mathbb{R}^n \times \mathbb{G}_N(\mathbb{R}^d) \to \mathbb{R}^n$$

such that, for $x_0 \in \mathbb{R}^n$, the solution $(X_t^{x_0})_{t \geq 0}$ of the SDE (2.1) can be written

$$X_t^{x_0} = F(x_0, B_t^*) + t^{\frac{N+1}{2}} \mathbf{R}_N(t),$$

where:

(1) $(B_t^*)_{t\geq 0}$ *is the lift of* $(B_t)_{t\geq 0}$ *in the group* $\mathbb{G}_N(\mathbb{R}^d)$;

(2) *The remainder term* $\mathbf{R}_N(t)$ *is bounded in probability when* $t \to 0$.

Proof. Let $N \geq 1$, and denote $\pi_j : \mathbb{R}^n \to \mathbb{R}$, $j = 1, ..., n$, the canonical projections defined by $\pi_j(x) = x_j$. By using iterations of Itô's formula and the scaling property of Brownian motion, we get

$$\left[\exp \left(\sum_{k=1}^{N} \sum_{I=(i_1,...,i_k)} \Lambda_I(B)_t V_I \right) \pi_j \right](x_0)$$

$$= \pi_j(x_0) + \sum_{k=1}^{N} \sum_{I=(i_1,...i_k)} (V_{i_1}...V_{i_k}\pi_j)(x_0) \int_{\Delta^k[0,t]} \circ dB^I + t^{\frac{N+1}{2}} \tilde{\mathbf{R}}_N^j(t)$$

$$= \pi_j(X_t^{x_0}) + t^{\frac{N+1}{2}} \hat{\mathbf{R}}_N^j(t),$$

where the remainder terms $\tilde{\mathbf{R}}_N^j$ and $\hat{\mathbf{R}}_N^j$ are bounded in probability when $t \to 0$. $\qquad \square$

Remark 2.11 *It is possible to show that more precisely,* $\exists\, \alpha, c > 0$ *such that* $\forall A > c$,

$$\lim_{t \to 0} \mathbb{P} \left(\sup_{0 \leq s \leq t} s^{\frac{N+1}{2}} \mid \mathbf{R}_N(s) \mid \geq At^{\frac{N+1}{2}} \right) \leq \exp \left(-\frac{A^\alpha}{c} \right).$$

For further details, we refer to [Castell (1993)].

2.5   An introduction to rough paths theory

As already pointed out, in general, the solution of the stochastic differential equation

$$X_t^{x_0} = x_0 + \sum_{i=1}^{d} \int_0^t V_i(X_s^{x_0}) \circ dB_s^i, \ 0 \leq t \leq T,$$

is not a continuous function of $(B_t)_{0 \leq t \leq T}$ for the topology of uniform convergence. For instance, it is easily seen that the *Levy area* functional

$$\left(\int_0^t B_s^1 dB_s^2 - B_s^2 dB_s^1 \right)_{0 \leq t \leq T}$$

is not continuous with respect to $(B_t^1, B_t^2)_{0 \leq t \leq T}$ (cf. [Watanabe (1984)]). Observe however, that it is a direct consequence of Section 2.3. that X^{x_0}

is continuous with respect to B as soon as the vector fields V_i's commute. This problem of the continuity of the Itô map

$$B \to X,$$

is solved in the setting of the rough paths theory which has recently been developed in [Lyons (1998)] (see also the survey [Lejay (2004)] and the book [Lyons and Qian (2002)]). Following N. Victoir, we believe that the free Carnot groups are a good framework for the rough paths theory. For $p \geq 1$, let us denote $\Omega^p \mathbb{G}_N(\mathbb{R}^d)$ the closure of the set of absolutely continuous horizontal paths $x^* : [0, T] \to \mathbb{G}_N(\mathbb{R}^d)$ with respect to the distance in p-variation which is given by

$$\delta_p(x^*, y^*) = \sup_\pi \left(\sum_{k=1}^{n-1} d_N \left(y_{t_i}^*(x_{t_i}^*)^{-1}, y_{t_{i+1}}^*(x_{t_{i+1}}^*)^{-1} \right)^p \right)^{\frac{1}{p}},$$

where the supremum is taken over all the subdivisions

$$\pi = \{ 0 \leq t_1 \leq \cdots \leq t_n \leq T \}$$

and where d_N denotes the Carnot-Carathéodory distance on the group $\mathbb{G}_N(\mathbb{R}^d)$. Consider now the map $\mathcal{I}$ which associates with an absolutely continuous path $x : [0, T] \to \mathbb{R}^d$ the absolutely continuous path $y : [0, T] \to \mathbb{R}^d$ that solves the ordinary differential equation

$$y_t = \sum_{i=1}^{d} \int_0^t V_i(y_s) dx_s^i.$$

It is clear that there exists a unique map $\mathcal{I}^*$ from the set of absolutely continuous horizontal paths $[0, T] \to \mathbb{G}_N(\mathbb{R}^d)$ onto the set of absolutely continuous horizontal paths $[0, T] \to \mathbb{G}_N(\mathbb{R}^n)$ which makes the following diagram commutative

$$
\begin{array}{ccc}
 & \mathcal{I}^* & \\
x^* & \longrightarrow & y^* \\
\uparrow & & \uparrow \\
x & \longrightarrow & y \\
 & \mathcal{I} &
\end{array}.
$$

The fundamental theorem of Lyons is the following:

Theorem 2.7 *If $N \geq [p]$, then in the topology of p-variation, there exists a continuous extension of $\mathcal{I}^*$ from $\Omega^p \mathbb{G}_N(\mathbb{R}^d)$ onto $\Omega^p \mathbb{G}_N(\mathbb{R}^n)$.*

One of the important consequences of this theorem, is that it enables to define the notion of solution for ordinary differential equations which are driven by paths x whose regularity is only $\frac{1}{p}$-Hölder. From Lyons theorem, to do so, we just have to find $y \in \Omega^p \mathbb{G}_N(\mathbb{R}^d)$ which satisfies $\pi y = x$, where π is the canonical projection $\mathbb{G}_N(\mathbb{R}^d) \to \mathbb{R}^d$ (finding such a lift, which is not unique in general, is always possible from [Lyons and Victoir (2004)]). In this direction let us mention the work [Coutin and Qian (2002)] who defined a notion of solution for stochastic differential equations driven by a fractional Brownian motion with Hurst parameter greater than $\frac{1}{4}$.

Observe that if $p < 2$, the Lyons fundamental theorem reduces to the classical theory of ordinary differential equations where the integrals are understood in Young's sense. The case $p \in (2,3)$ is of particular importance since it covers the theory of stochastic differential equations driven by Brownian paths (a Brownian path is almost surely $\frac{1}{2} - \varepsilon$ Hölder continuous, $\varepsilon > 0$). This point of view on stochastic differential equations is for instance used in [Friz and Victoir (2003)] and [Ledoux *et al.* (2002),] to obtain new and simplified proofs of Stroock-Varadhan support theorem and of the Freidlin-Wentzell theory.

Chapter 3

Hypoelliptic Flows

Chapter 1 has shown that, thanks to the Chen-Strichartz development the Itô's map

$$\mathcal{I} : \mathcal{C}\left([0,T], \mathbb{R}^d\right) \to \mathcal{C}\left([0,T], \mathbb{R}^n\right), \ B \to X,$$

established by the stochastic differential equation

$$X_t^{x_0} = x_0 + \sum_{i=1}^{d} \int_0^t V_i(X_s^{x_0}) \circ dB_s^i,$$

could be *formally* factorized in the following manner

$$\mathcal{I} = F \circ H.$$

The map

$$H : \mathcal{C}\left([0,T], \mathbb{R}^d\right) \to \mathcal{C}\left([0,T], \exp\left(\mathfrak{g}_{d,\infty}\right)\right),$$

is an horizontal lift in $\exp\left(\mathfrak{g}_{d,\infty}\right)$ where $\mathfrak{g}_{d,\infty}$ is the free Lie algebra with d generators. And F is simply a map $\exp\left(\mathfrak{g}_{d,\infty}\right) \to \mathbb{R}^n$. In Chapter 2 this factorization was made totally rigorous in the case where the vector fields V_i's generate a nilpotent Lie algebra. In this chapter we now go one step further by studying the case where the V_i's do not generate a nilpotent Lie algebra anymore but satisfy the so-called strong Hörmander's condition. In that case, it shall be shown that the Gromov's notion of tangent space can be used to approximate locally the geometry of the Itô's map by the geometry of a Carnot group.

We start the chapter with the probabilistic proof of Hörmander's theorem. This proof is based on a stochastic calculus of variations that has been developed by Paul Malliavin (cf. [Malliavin (1978)]). The idea of the proof is to show that the Itô's map $\mathcal{I}$ is differentiable in a weak sense and then to show that, under Hörmander's conditions, the derivative is non degenerate.

Then, we introduce the basic background of differential geometry that is needed to study locally the stochastic flow associated with a hypoelliptic differential system. This local study of the flow is used to derive the behaviour in small times of a hypoelliptic kernel on the diagonal.

After, we provide a large class of examples of hypoelliptic operators which arise naturally in a manifold context. These are the horizontal Laplacians on principal bundles, whose hypoellipticity is related to non vanishing curvature conditions. The case of the horizontal diffusion over a Riemannian manifold is investigated in details.

The local behaviour of the hypoelliptic heat kernel on the diagonal is then used to get informations on the spectrum of hypoelliptic operators defined on compact manifolds.

Finally, we conclude the chapter by the study of stochastic flows associated with stochastic differential equations driven by loops.

3.1 Hypoelliptic operators and Hörmander's theorem

Consider a stochastic differential equation

$$X_t^{x_0} = x_0 + \sum_{i=1}^{d} \int_0^t V_i(X_s^{x_0}) \circ dB_s^i, \ t \geq 0, \tag{3.1}$$

where $x_0 \in \mathbb{R}^n$, $V_1, ..., V_d$ are C^∞ bounded vector fields on $\mathbb{R}^n$ and $(B_t)_{t \geq 0}$ is a d-dimensional standard Brownian motion.

Let us first recall (see Appendix A) that for every $x_0 \in \mathbb{R}^n$ and every smooth function $f : \mathbb{R}^n \to \mathbb{R}$ which is compactly supported,

$$\mathbb{E}\left(f(X_t^{x_0})\right) = \left(e^{t\mathcal{L}} f\right)(x_0), \tag{3.2}$$

where

$$\mathcal{L} = \sum_{i=1}^{d} V_i^2.$$

In this section we shall be interested in the existence of smooth densities for the random variables $X_t^{x_0}$, $t > 0$, $x_0 \in \mathbb{R}^n$. According to formula (3.2), this question is therefore equivalent to the question of the existence of a smooth transition kernel with respect to the Lebesgue measure for the operators

$e^{t\mathcal{L}}$. Let us recall the following definition which comes from functional analysis.

Definition 3.1 A differential operator $\mathcal{G}$ defined on an open set $\mathcal{O} \subset \mathbb{R}^n$ is called hypoelliptic if, whenever u is a distribution on $\mathcal{O}$, u is a smooth function on any open set $\mathcal{O}' \subset \mathcal{O}$ on which $\mathcal{G}u$ is smooth.

It is possible to show that the existence of a smooth transition kernel with respect to the Lebesgue measure for $e^{t\mathcal{L}}$ is equivalent to the hypoellipticity of $\mathcal{L}$. Therefore, our initial question about the existence of smooth densities for the random variables $X_t^{x_0}$, $t > 0$, $x_0 \in \mathbb{R}^n$ is reduced to the study of the hypoellipticity of $\mathcal{L}$.

Let us denote by $\mathfrak{L}$ the Lie algebra generated by the vector fields V_i's and for $p \geq 2$, by $\mathfrak{L}^p$ the Lie subalgebra inductively defined by

$$\mathfrak{L}^p = \{[X, Y], \ X \in \mathfrak{L}^{p-1}, Y \in \mathfrak{L}\}.$$

Moreover if $\mathfrak{a}$ is a subset of $\mathfrak{L}$, we denote

$$\mathfrak{a}(x) = \{V(x), V \in \mathfrak{a}\}, \ x \in \mathbb{R}^n.$$

The celebrated Hörmander's theorem proved in [Hörmander (1967)] is the following:

Theorem 3.1 *Assume that for every $x_0 \in \mathbb{R}^n$,*

$$\mathfrak{L}(x_0) = \mathbb{R}^n,$$

then the operator $\mathcal{L}$ is hypoelliptic.

Remark 3.1 *This is also a necessary condition when the V_i's are with analytic coefficients (see [Derridj (1971)]).*

Remark 3.2 *It is also possible to obtain hypoellipticity results for second order differential operators which can not be written as a sum of squares (see [Oleinic and Radkevic (1973)]).*

The original proof of Hörmander was rather complicated and has been considerably simplified in [Kohn (1973)] using the theory of pseudo-differential operators. The probabilistic counterpart of the theorem, which is the existence of a smooth density for the random variable $X_t^{x_0}$, $t > 0$, has first been pointed out in [Malliavin (1978)] where, in order the reprove the theorem under weaker assumptions, the author has developed a stochastic calculus of variations which is now known as the Malliavin calculus (see Appendix A and [Nualart (1995)] for a very complete exposition). After Malliavin's work, let us mention the work [Bismut (1981)] in which the author uses interesting integration by parts formulae to prove the theorem. This is this probabilistic counterpart that we shall now prove.

Theorem 3.2 *Assume that at some $x_0 \in \mathbb{R}^n$ we have*

$$\mathcal{L}(x_0) = \mathbb{R}^n, \tag{3.3}$$

then, for any $t > 0$, the random variable $X_t^{x_0}$ has a smooth density with respect to the Lebesgue measure of $\mathbb{R}^n$, where $(X_t^{x_0})_{t \geq 0}$ denotes the solution of (3.1).

Proof. We shall show that $X_1^{x_0}$ admits a density with respect to the Lebesgue measure, by using the Malliavin covariance matrix Γ (see Appendix A) associated with $X_1^{x_0}$. Observe, that for notational convenience we take $t = 1$, but the proof is exactly the same for any $t > 0$. We proceed in several steps. In a first step, we perform the computation of Γ, in a second step we show that Γ is almost surely invertible and finally, we shall qualitatively explain why for every $p > 1$,

$$\mathbb{E}\left(\frac{1}{|\det \Gamma|^p}\right) < +\infty.$$

Step 1. By definition, we have

$$\Gamma = \left(\int_0^1 \langle \mathbf{D}_s X_1^i, \mathbf{D}_s X^j \rangle_{\mathbb{R}^d} ds\right)_{1 \leq i,j \leq n},$$

but from Theorem (A.7) of Appendix A,

$$\mathbf{D}_t^j X_1 = \mathbf{J}_{0 \to 1} \mathbf{J}_{0 \to t}^{-1} V_j(X_t), \quad j = 1, ..., d, \quad 0 \leq t \leq 1,$$

where $(\mathbf{J}_{0 \to t})_{t \geq 0}$ is the first variation process defined by

$$\mathbf{J}_{0 \to t} = \frac{\partial X_t^x}{\partial x}.$$

Therefore,

$$\Gamma = \mathbf{J}_{0 \to 1} \int_0^1 \mathbf{J}_{0 \to t}^{-1} V(X_t)^T V(X_t)^T \mathbf{J}_{0 \to t}^{-1} dt \,{}^T\mathbf{J}_{0 \to 1},$$

where V denotes the $n \times d$ matrix $(V_1 ... V_d)$.

Step 2. Since $J_{0\to1}$ is almost surely invertible, in order to show that Γ is invertible with probability one, it is enough to check that with probability one, the matrix

$$C = \int_0^1 J_{0\to t}^{-1} V(X_t)^T V(X_t)^T J_{0\to t}^{-1} dt$$

is invertible. For this, let us introduce the family of processes

$$C_t = \int_0^t J_{0\to s}^{-1} V(X_s)^T V(X_s)^T J_{0\to s}^{-1} dt, \quad t \ge 0.$$

We denote the kernel of C_t by $K_t \subset \mathbb{R}^n$ and get a decreasing sequence of random subspaces of $\mathbb{R}^n$. From Blumenthal zero-one law, the space $V = \cup_{t>0} K_t$ is deterministic with probability one.

Let now $y \in V$, and consider the stopping time

$$\theta := \inf\{s > 0, \; {}^T y C_s y > 0\}$$

Then $\theta > 0$ almost surely. From the expression of C_t, we get therefore that for $0 \le s < \theta$,

$$^T y \; J_{0\to s}^{-1} V_i(X_s) = 0, \quad i = 1, ..., d.$$

By applying this at $s = 0$, we obtain first

$$^T y \; V_i(x_0) = 0, \quad i = 1, ..., d.$$

Observe now that

$$J_{0\to s}^{-1} V_i(X_s) = \Phi_s^* V_i,$$

where Φ denotes the stochastic flow associated with equation (3.1), and where $\Phi_s^* V_i$ denotes the pull-back action of Φ on V_i. Therefore, according to the Itô's formula given in Proposition A.6 of Appendix A, we obtain that for $t < \theta$,

$$\sum_{j=1}^{d} \int_0^t \left(\Phi_s^* \mathcal{L}_{V_j} V\right)(x_0) \circ dB_s^j = 0, \quad i = 1, ..., d,$$

that is

$$\sum_{j=1}^{d} \int_0^t \, {}^T y \; (\Phi_s^*[V_j, V_i])(x_0) \circ dB_s^j = 0, \quad i = 1, ..., d.$$

Therefore, due to the uniqueness of the semimartingale decomposition, for $0 \leq s < \theta$,

$$^Ty \left(\Phi_s^*[V_j, V_i] \right)(x_0), \quad i, j = 1, ..., d.$$

By applying this at $s = 0$, we obtain then

$$^Ty \, [V_j, V_i](x_0) = 0, \quad i, j = 1, ..., d.$$

An iteration of the Itô's formula given in Proposition A.6 of Appendix A shows then that, we actually have

$$^Ty \, U(x_0) = 0, \quad U \in \mathfrak{L}(x_0),$$

so that $y = 0$. From this, we conclude that $V = 0$ and thus that C is invertible with probability one. Therefore, the random variable $X_1^{x_0}$ admits a density with respect to the Lebesgue measure.

Step 3. To show that the density of $X_1^{x_0}$ is smooth, we have to show that for every $p > 1$,

$$\mathbb{E} \left(\frac{1}{|\det \Gamma|^p} \right) < +\infty.$$

Recall now that, with the previous notations

$$\Gamma = \mathbf{J}_{0 \to 1} C \, ^T \mathbf{J}_{0 \to 1}.$$

By differentiating the stochastic differential equation (3.1) with respect to the initial condition, we obtain (non-autonomous linear equations for the processes $(\mathbf{J}_{0 \to t})_{t \geq 0}$ and $(\mathbf{J}_{0 \to t}^{-1})_{t \geq 0}$; see [Malliavin (1997)], pp. 240, for further details. These equations make easy the proof of the fact that for every $p > 1$,

$$\mathbb{E} \left(\frac{1}{|\det \mathbf{J}_{0 \to 1}|^{2p}} \right) < +\infty.$$

Therefore, it remains to show that for every $p > 1$,

$$\mathbb{E} \left(\frac{1}{|\det C|^p} \right) < +\infty.$$

The idea is to introduce again the process

$$C_t = \int_0^t \mathbf{J}_{0 \to s}^{-1} V(X_s)^T V(X_s)^T \mathbf{J}_{0 \to s}^{-1} dt, \quad t \geq 0,$$

and to control the smallest eigenvalue of C_t by showing that it can not be too small. It can be done by estimating the quantity

$$\sup_{\|v\|=1} \mathbb{P}\left({}^T\!v\,C_t v \leq \varepsilon\right) = \sup_{\|v\|=1} \mathbb{P}\left(\sum_{j=1}^{d} \int_0^t \left({}^T\!v\,\mathbf{J}_{0\rightarrow s}^{-1} V_j(X_s)\right)^2 ds \leq \varepsilon\right)$$

with the help of the so-called Norris lemma. This lemma roughly says that, when the quadratic variation or the bounded variation part of a continuous semimartingale is large, then the semimartingale can only be small with an exponentially small probability.

We do not go into details on this part, since it is very technical, and we believe that the most interesting aspect of the probabilistic proof of Hörmander's theorem is to see how $\mathcal{L}(x_0)$ appears in the Malliavin matrix. For further details on the Norris lemma and its use, we refer to [Nualart (1995)] pp. 116-123. $\quad\square$

Remark 3.3 *Therefore, under the assumption (3.3), there exists a smooth function p on $(0, +\infty) \times \mathbb{R}^n$ such that for any smooth function $f : \mathbb{R}^n \rightarrow \mathbb{R}$ which is compactly supported*

$$(P_t f)(x_0) = \int_{\mathbb{R}^n} p(t, y) f(y) dy,$$

where $(P_t)_{t\geq 0}$ denotes the transition function of the strong Markov process $(X_t^{x_0})_{t\geq 0}$. Moreover, the function p satisfies Kolmogorov's forward equation:

$$\frac{\partial p}{\partial t} = \mathcal{L}^* p,$$

where $\mathcal{L}^*$ is the adjoint of $\mathcal{L}$.

Remark 3.4 *There is a special case of Hörmander's theorem, which covers very many of the applications one meets in practice; this is the case where $\mathcal{L}$ is elliptic at x_0, that is $(V_1(x_0), ..., V_d(x_0))$ are enough to span $\mathbb{R}^n$. This special case had earlier been obtained by Hermann Weyl.*

Remark 3.5 *Let us mention that it possible to extend Hörmander's theorem to stochastic differential equations valued in a separable Hilbert space of the type*

$$dX_t = (AX_t + \alpha(X_t))dt + \sum_{i=1}^{d} \sigma_i(X_t) dB_t^i,$$

where A generates a strongly continuous group; for further details we refer to [Baudoin and Teichmann (2003)].

One can extend slightly Hörmander's theorem.

Theorem 3.3   *Let $x_0 \in \mathbb{R}^n$ and $N \in \mathbb{N}$. If $\mathfrak{L}^{N+1}(x_0) = \mathbb{R}^n$, then for any $t > 0$, the random variable*

$$(X_t^{x_0}, B_t^*)$$

has a smooth density with respect to the Lebesgue measure of $\mathbb{R}^n \times \mathbb{G}_N(\mathbb{R}^d)$, where $(X_t^{x_0})_{t \geq 0}$ is the solution of (3.1) with initial condition x_0 and $(B_t^*)_{t \geq 0}$ the lift of $(B_t)_{t \geq 0}$ in the free Carnot group $\mathbb{G}_N(\mathbb{R}^d)$.

Proof.   With a slight abuse of notation, we still denote V_i (resp. D_i) the extension of V_i (resp. D_i) to the space $\mathbb{R}^n \times \mathbb{G}_N(\mathbb{R}^d)$. The process $(X_t^{x_0}, B_t^*)_{t \geq 0}$ is easily seen to be a diffusion process in $\mathbb{R}^n \times \mathbb{G}_N(\mathbb{R}^d)$ with infinitesimal generator

$$\frac{1}{2} \sum_{i=1}^d (V_i + D_i)^2.$$

Thus, to prove the theorem, it is enough to check the Hörmander's condition for this operator at the point $(x_0, 0)$. Now, notice that for $1 \leq i, j \leq n$, $[V_i, D_j] = 0$, so that

$$\mathbf{Lie}(V_1 + D_1, ..., V_n + D_n)(x_0, 0) \simeq \mathfrak{L}^{N+1}(x_0) \oplus \mathfrak{g}_N(\mathbb{R}^d),$$

because $\mathfrak{g}_N(\mathbb{R}^d)$ is step N nilpotent. We denoted $\mathbf{Lie}(V_1 + D_1, ..., V_n + D_n)$ the Lie algebra generated by $(V_1 + D_1, ..., V_n + D_n)$. The conclusion follows readily. $\square$

Example 3.1   For $N = 0$, we have $\mathbb{G}_0(\mathbb{R}^d) = \{0\}$ and Theorem 3.3 is the classical Hörmander's theorem.

Example 3.2   For $N = 1$, we have $\mathbb{G}_1(\mathbb{R}^d) \simeq \mathbb{R}^d$ and Theorem 3.3 gives a suffcient condition for the existence of a smooth density for the variable

$$(X_t^{x_0}, B_t).$$

Example 3.3   For $N = 2$, we have $\mathbb{G}_2(\mathbb{R}^d) \simeq \mathbb{R}^d \times \mathbb{R}^{\frac{d(d-1)}{2}}$ and Theorem 3.3 gives a suffcient condition for the existence of a smooth density for the variable

$$(X_t^{x_0}, B_t, \wedge B_t).$$

where

$$\wedge B_t = \left(\frac{1}{2} \int_0^t B_s^i dB_s^j - B_s^j dB_s^i \right)_{1 \le i < j \le d}.$$

3.2 Sub-Riemannian geometry

The goal of this section is to present the basic results of sub-Riemannian geometry, which is the natural geometry associated with hypoelliptic operators. We essentially aim at generalizing the geometry of the Carnot Groups that has been presented in the previous chapter.

From now on, we consider d vector fields $V_i : \mathbb{R}^n \to \mathbb{R}^n$ which are C^∞ bounded and shall always assume that the following assumption is satisfied.

Strong Hörmander's Condition: *For every* $x \in \mathbb{R}^n$, *we have*:

$$\text{span}\{V_I(x), I \in \cup_{k \ge 1}\{1, ..., d\}^k\} = \mathbb{R}^n.$$

We recall that if $I = (i_1, ..., i_k) \in \{1, ..., d\}^k$ is a word, we denote by V_I the commutator defined by

$$V_I = [V_{i_1}, [V_{i_2}, ..., [V_{i_{k-1}}, V_{i_k}]...].$$

Let us mention that in sub-Riemannian litterature, the strong Hörmander's condition is more often called the Chow's condition or bracket generating's condition.

Definition 3.2 The set of linear combinations with smooth coefficients of the vector fields $V_1, ..., V_d$ is called the differential system (or sheaf) generated by these vector fields. It shall be denoted in the sequel $\mathcal{D}$.

With this definition, we can roughly say that the goal of the sub-Riemannian geometry is to study the intrinsic object of the distribution, such the symmetry groups, the local differential invariants, etc...

Observe that $\mathcal{D}$ is naturally endowed with a structure of $C_\infty(\mathbb{R}^n, \mathbb{R})$-module. For $x \in \mathbb{R}^n$, we denote

$$\mathcal{D}(x) = \{X(x), X \in \mathcal{D}\}.$$

If the integer $\dim \mathcal{D}(x)$ does not depend on x, then $\mathcal{D}$ is said to be a distribution. Observe that the Lie bracket of two distributions is not necessarily a distribution, so that we really have to work with differential systems. In the case where $\dim \mathcal{D}(x)$ is constant and equals the dimension of the ambient space, that is n, the distribution is said to be elliptic. The study of elliptic distributions is exactly the Riemannian geometry. The sub-Riemannian geometry is much more *rough* than the Riemannian one. Some of the major differences between the two geometries are the following:

(1) In general, it does not exist a canonical connection like the Levi-Civita connection of Riemannian geometry;
(2) The Hausdorff dimension is greater than the manifold dimension (see Chapter 2, in the case of the Carnot groups);
(3) The exponential map is never a local diffeomorphism in a neighborhood of the point at which it is based (see [Rayner (1967)]);
(4) The space of paths tangent to the differential system and joining two fixed points can have singularities (see [Montgomery (2002)]).

Before we go into the heart of the subject, it is maybe useful to provide some examples.

Example 3.4   Let $\mathbb{G}$ be a Carnot group and consider for $\mathcal{D}$ the left invariant differential system which is generated by the basis of $\mathbb{G}$. Then, $\mathcal{D}$ satisfies the strong Hörmander's condition and the sub-Riemannian geometry associated with $\mathcal{D}$ is precisely the geometry that has been studied in Chapter 2.

Example 3.5   Let $\mathbb{M}$ be a manifold of dimension d. Assume that there exists on $\mathbb{M}$ a family of vector fields $(V_1, ..., V_d)$ such that for every $x \in \mathbb{M}$, $(V_1(x), ..., V_d(x))$ is a basis of the tangent space at x . Let us denote by $\mathcal{D}$ the differential system generated by $(V_1, ..., V_d)$ (it is actually a distribution). Then, $\mathcal{D}$ satisfies the strong Hörmander's condition and the sub-Riemannian geometry associated with $\mathcal{D}$ is the Riemannian geometry on $\mathbb{M}$ which is induced by the moving frame $(V_1(x), ..., V_d(x))$.

Example 3.6   Let us consider the Lie group $\mathbf{SO}(3)$, i.e. the group of 3×3, real, orthogonal matrices of determinant 1. Its Lie algebra $\mathfrak{so}(3)$ consists of 3×3, real, skew-adjoint matrices of trace 0. A basis of $\mathfrak{so}(3)$ is formed by

$$V_1 = \begin{pmatrix} 0 & 1 & 0 \\ -1 & 0 & 0 \\ 0 & 0 & 0 \end{pmatrix}, \ V_2 = \begin{pmatrix} 0 & 0 & 0 \\ 0 & 0 & 1 \\ 0 & -1 & 0 \end{pmatrix}, \ V_3 = \begin{pmatrix} 0 & 0 & 1 \\ 0 & 0 & 0 \\ -1 & 0 & 0 \end{pmatrix}$$

Observe that the following commutation relations hold

$$[V_1, V_2] = V_3, \ [V_2, V_3] = V_1, \ [V_3, V_1] = V_2,$$

so that the differential system $\mathcal{D}$ which is generated by V_1 and V_2 satisfies the strong Hörmander's condition. The group $\mathbf{SO}(3)$ can be seen as the orthonormal frame bundle of the unit sphere $\mathbb{S}^2$ and, via this identification, $\mathcal{D}$ is generated by the horizontal lifts of vector fields on $\mathbb{S}^2$. Therefore, in a way, the sub-Riemannian geometry associated with $\mathcal{D}$ is the geometry of the holonomy of $\mathbb{S}^2$. This example shall be widely generalized at the end of this chapter; actually many interesting examples of sub-Riemannian geometries arise from principal bundles.

Example 3.7 Let us consider the Lie group $\mathbf{SU}(2)$, i.e. the group of 2×2, complex, unitary matrices of determinant 1. Its Lie algebra $\mathfrak{su}(2)$ consists of 2×2, complex, skew-adjoint matrices of trace 0. A basis of $\mathfrak{su}(2)$ is formed by

$$V_1 = \frac{1}{2} \begin{pmatrix} i & 0 \\ 0 & -i \end{pmatrix}, \ V_2 = \frac{1}{2} \begin{pmatrix} 0 & 1 \\ -1 & 0 \end{pmatrix}, \ V_3 = \frac{1}{2} \begin{pmatrix} 0 & i \\ i & 0 \end{pmatrix}.$$

Note the commutation relations

$$[V_1, V_2] = V_3, \ [V_2, V_3] = V_1, \ [V_3, V_1] = V_2,$$

so that the differential system $\mathcal{D}$ which is generated by V_1 and V_2 satisfies the strong Hörmander's condition. Let us mention that there exists an explicit homomorphism $\mathbf{SU}(2) \to \mathbf{SO}(3)$ which exhibits $\mathbf{SU}(2)$ as a double cover of $\mathbf{SO}(3)$, so that this example is actually a consequence of the previous one.

Definition 3.3 An absolutely continuous path $c : [0,1] \to \mathbb{R}^n$ is said to be horizontal (with respect to $\mathcal{D}$) if for almost every $s \in [0,1]$,

$$c'(s) \in \mathcal{D}(c(s)).$$

We can now give the Chow's theorem in its full generality.

Theorem 3.4 *Let $(x, y) \in \mathbb{R}^n \times \mathbb{R}^n$. There exists at least one absolutely continuous horizontal curve $c : [0,1] \to \mathbb{R}^n$ such that $c(0) = x$ and $c(1) = y$.*

Unlike the case of a Carnot group, the proof of this theorem is far to be easy (see the Chapter 2 of [Montgomery (2002)] for a complete and detailed proof). Thanks to this theorem, it is possible like we did it in the case of the Carnot groups, to define a distance on $\mathbb{R}^n$. From the definition of $\mathcal{D}$, any absolutely continuous and horizontal curve $c : [0,1] \to \mathbb{R}^n$ satisfies

$$c'(s) = \sum_{i=1}^{d} c_i'(s) V_i(c(s)),$$

for some absolutely continuous curves $c_i : [0,1] \to \mathbb{R}$. The length of c is defined by

$$l(c) = \int_0^1 \sqrt{\sum_{i=1}^{d} c_i'(s)^2} ds.$$

The Carnot-Carathéodory distance between two points a and b in $\mathbb{R}^n$ is now defined as being

$$d(a,b) = \inf l(c),$$

where the infimum is taken over the set of absolutely continuous horizontal curves $c : [0,1] \to \mathbb{R}^n$ such that

$$c(0) = a, \ c(1) = b.$$

It can be shown, by proving a generalization in this framework of the ball-box theorem seen in Chapter 2, that the topology given by the Carnot-Carathéodory distance d is the usual Euclidean topology of $\mathbb{R}^n$.

A horizontal curve with length $d(a,b)$ is called a sub-Riemannian geodesic joining a and b. With Chow's theorem in mind, we address now the problem of existence of geodesics, whose Riemannian analogue is the classical Hopf-Rinow theorem.

Theorem 3.5 *Any two points of $\mathbb{R}^n$ can be joined by a geodesic.*

There is a major difference between Riemannian geodesics and sub-Riemannian ones: unlike the two step Carnot group case, it may happen that a geodesic is not smooth.

3.3   The tangent space to a hypoelliptic diffusion

First, we have to introduce some concepts of differential geometry. The set of linear combinations with smooth coefficients of the vector fields $V_1, ..., V_d$ is called the differential system (or sheaf) generated by these vector fields.

It shall be denoted in the sequel $\mathcal{D}$. Notice that $\mathcal{D}$ is naturally endowed with a structure of $C_\infty(\mathbb{R}^n, \mathbb{R})$-module. For $x \in \mathbb{R}^n$, we denote

$$\mathcal{D}(x) = \{X(x), X \in \mathcal{D}\}.$$

If the integer $\dim \mathcal{D}(x)$ does not depend on x, then $\mathcal{D}$ is said to be a distribution. Observe that the Lie bracket of two distributions is not necessarily a distribution, so that we really have to work with differential systems. The Lie brackets of vector fields in $\mathcal{V}$ generates a flag of differential systems,

$$\mathcal{D} \subset \mathcal{D}^2 \subset \cdots \subset \mathcal{D}^k \subset \cdots,$$

where $\mathcal{D}^k$ is recursively defined by the formula

$$\mathcal{D}^k = \mathcal{D}^{k-1} + [\mathcal{D}, \mathcal{D}^{k-1}].$$

As a module, $\mathcal{D}^k$ is generated by the set of vector fields V_I, where I describes the set of words with length k. Moreover, due to Jacobi identity, we have $[\mathcal{D}^i, \mathcal{D}^j] \subset \mathcal{D}^{i+j}$. This flag is called the *canonical flag* associated with the differential system $\mathcal{D}$. Hörmander's strong condition, which we supposed to hold, states that for each $x \in \mathbb{R}^n$, there is a smallest integer $r(x)$ such that $\mathcal{D}^{r(x)} = \mathbb{R}^n$. This integer is called the degree of non holonomy at x. Notice that r is upper continuous function, that is $r(y) \le r(x)$ for y near x. For each $x \in \mathbb{R}^n$, the canonical flag induces a flag of vector subspaces,

$$\mathcal{D}(x) \subset \mathcal{D}^2(x) \subset \cdots \subset \mathcal{D}^{r(x)}(x) = \mathbb{R}^n.$$

The integer list $\left(\dim \mathcal{D}^k(x)\right)_{1 \le k \le r(x)}$ is called the growth vector of $\mathcal{V}$ at x. The point x is said to be a regular point of $\mathcal{V}$ if the growth vector is constant in a neighborhood of x. Otherwise, we say that x is a singular point. On a Carnot group, due to the homogeneity, all points are regular.

We are now able to define in a purely algebraic manner what shall be relevant for us: the nilpotentization and the tangent space of $\mathcal{V}$ at a regular point. Later, we shall see that this tangent space also can be constructed in a purely metric manner. Let $\mathcal{V}_i = \mathcal{D}^i/\mathcal{D}^{i-1}$ denote the quotient differential systems, and define

$$\mathcal{N}(\mathcal{D}) = \mathcal{V}_1 \oplus \cdots \oplus \mathcal{V}_k \oplus \cdots.$$

The Lie bracket of vector fields induces a bilinear map on $\mathcal{N}(\mathcal{D})$ which respects the grading: $[\mathcal{V}_i, \mathcal{V}_j] \subset \mathcal{V}_{i+j}$. Actually, $\mathcal{N}(\mathcal{D})$ inherits the structure of a sheaf of Lie algebras. Moreover, if x is a regular point of $\mathcal{D}$, then this bracket induces a $r(x)$-step nilpotent graded Lie algebra structure on

$\mathcal{N}(\mathcal{D})(x)$. Observe that the dimension of $\mathcal{N}(\mathcal{D})(x)$ is equal to n and that from the definition, $(V_1(x), ..., V_d(x))$ Lie generates $\mathcal{N}(\mathcal{D})(x)$.

Definition 3.4   If x is a regular point of $\mathcal{D}$, the $r(x)$-step nilpotent graded Lie algebra $\mathcal{N}(\mathcal{D})(x)$ is called the nilpotentization of $\mathcal{D}$ at x. This Lie algebra is the Lie algebra of a unique Carnot group which shall be denoted $\mathbf{Gr}(\mathcal{D})(x)$ and called the tangent space to $\mathcal{D}$ at x.

Remark 3.6   *At a regular point x, $\mathbf{Gr}(\mathcal{D})(x)$ can be seen as a quotient of $\mathbb{G}_{r(x)}(\mathbb{R}^d)$.*

Remark 3.7   *The notation $\mathbf{Gr}$ is for Gromov.*

Definition 3.5   If x is a regular point of $\mathcal{D}$, we say that x is a normal point of $\mathcal{D}$ if there exists a neighborhood U of x such that:

(1) for every $y \in U$, y is a regular point of $\mathcal{D}$;
(2) for every $y \in U$, there exists a Carnot algebra isomorphism

$$\psi : \mathcal{N}(\mathcal{D})(x) \to \mathcal{N}(\mathcal{D})(y),$$

such that $\psi(V_i(x)) = V_i(y)$, $i = 1, ..., d$.

Remark 3.8   *To say that x is a normal point of $\mathcal{D}$ is equivalent to say that any relation of the type*

$$\sum_{k=1}^{r(x)} \sum_{I=(i_1,...,i_k)} a_I V_I(x), \quad a_I \in \mathbb{R},$$

can be extended in a smooth way to a relation,

$$\sum_{k=1}^{r(x)} \sum_{I=(i_1,...,i_k)} a_I(y) V_I(y), \quad y \in U.$$

More roughly speaking, if x is a normal point of $\mathcal{D}$ the isomorphism class of $\mathcal{N}(\mathcal{D})(x)$ is constant in a neighborhood of x. Let us mention that in typical cases there exist regular points which are not normal (see [Gershkovich and Vershik (1988)]).

To illustrate these notions, we give now the nilpotentization and the tangent space in the examples already seen.

Example 3.8   Let $\mathbb{G}$ be a Carnot group with Lie algebra $\mathfrak{g}$ and consider for $\mathcal{D}$ the left invariant differential system which is generated by the basis of $\mathbb{G}$. Then, each point is normal and it is immediate that for every $x \in \mathbb{G}$,

$$\mathcal{N}(\mathcal{D})(x) = \mathfrak{g},$$

$$\mathbf{Gr}(\mathcal{D})(x) = \mathbb{G}.$$

Example 3.9 Let $\mathbb{M}$ be a manifold of dimension d. Assume that there exists on $\mathbb{M}$ a family of vector fields $(V_1, ..., V_d)$ such that for every $x \in \mathbb{M}$, $(V_1(x), ..., V_d(x))$ is a basis of the tangent space at x and denote by $\mathcal{D}$ the differential system generated by $(V_1, ..., V_d)$. Then, each point is normal and for every $x \in \mathbb{M}$,

$$\mathcal{N}(\mathcal{D})(x) = \mathbb{R}^d,$$

$$\mathbf{Gr}(\mathcal{D})(x) = \mathbb{R}^d.$$

Example 3.10 Let us consider on $\mathbf{SO}(3)$ the left invariant differential system $\mathcal{D}$ generated by

$$V_1 = \begin{pmatrix} 0 & 1 & 0 \\ -1 & 0 & 0 \\ 0 & 0 & 0 \end{pmatrix} \text{ and } V_2 = \begin{pmatrix} 0 & 0 & 0 \\ 0 & 0 & 1 \\ 0 & -1 & 0 \end{pmatrix}.$$

In that case, it is easily checked that each point is normal and that for every $x \in \mathbf{SO}(3)$,

$$\mathcal{N}(\mathcal{D})(x) = \mathfrak{g}_2(\mathbb{R}^2),$$

$$\mathbf{Gr}(\mathcal{D})(x) = \mathbb{G}_2(\mathbb{R}^2).$$

Example 3.11 Let us consider on $\mathbf{SU}(2)$ the left invariant distribution $\mathcal{D}$ which is generated by

$$V_1 = \frac{1}{2} \begin{pmatrix} i & 0 \\ 0 & -i \end{pmatrix} \text{ and } V_2 = \frac{1}{2} \begin{pmatrix} 0 & 1 \\ -1 & 0 \end{pmatrix}.$$

Each point is normal and for every $x \in \mathbf{SU}(2)$,

$$\mathcal{N}(\mathcal{D})(x) = \mathfrak{g}_2(\mathbb{R}^2),$$

$$\mathbf{Gr}(\mathcal{D})(x) = \mathbb{G}_2(\mathbb{R}^2).$$

A really striking fact is that the tangent space $\mathbf{Gr}(\mathcal{D})(x)$ at a regular point is not only a differential invariant but also a purely *metric invariant*. Actually Gromov discovered that it is possible, in a very general way, to define a notion of tangent space to an abstract metric space. This point of view is widely developed in [Gromov (1996)] and [Gromov (1999)], and is the starting point of the so-called metric geometry. The Gromov-Hausdorff distance between two metric spaces $\mathbb{M}_1$ and $\mathbb{M}_2$ is defined as follows: $\delta_{\mathbf{GH}}(\mathbb{M}_1, \mathbb{M}_2)$ is the infimum of real numbers ρ for which there exists isometric embeddings of $\mathbb{M}_1$ and $\mathbb{M}_2$ in a same metric space $\mathbb{M}_3$, say $i_1 : \mathbb{M}_1 \to \mathbb{M}_3$ and $i_2 : \mathbb{M}_2 \to \mathbb{M}_3$, such that the Hausdorff distance of $i_1(\mathbb{M}_1)$ and $i_2(\mathbb{M}_2)$ as subsets of $\mathbb{M}_3$ is lower than ρ. The important property of the Gromov-Hausdorff distance is the following theorem.

Theorem 3.6 *Suppose $\mathbb{M}_1$ and $\mathbb{M}_2$ are complete metric spaces with $\delta_{\mathbf{GH}}(\mathbb{M}_1, \mathbb{M}_2) = 0$. Then $\mathbb{M}_1$ and $\mathbb{M}_2$ are isometric.*

Thanks to this distance, we have now a convenient notion of limit of a sequence metric spaces.

Definition 3.6 A sequence of pointed metric spaces $(\mathbb{M}_n, x_n)$ is said to Gromov-Hausdorff converge to the pointed metric space $(\mathbb{M}, x)$ if for any positive R

$$\lim_{n \to +\infty} \delta_{\mathbf{GH}} \left(\mathbf{B}_{\mathbb{M}_n} (x_n, R), \mathbf{B}_{\mathbb{M}} (x, R) \right) = 0,$$

where $\mathbf{B}_{\mathbb{M}_n} (x_n, R)$ is the open ball centered at x_n with radius R in $\mathbb{M}_n$. In that case we shall write

$$\lim_{n \to +\infty} (\mathbb{M}_n, x_n) = (\mathbb{M}, x).$$

If $\mathbb{M}$ is a metric space and $\lambda > 0$, we denote $\lambda \cdot \mathbb{M}$ the new metric space obtained by multiplying all distances by λ.

Definition 3.7 Let $\mathbb{M}$ be a metric space and $x_0 \in \mathbb{M}$. If the Gromov-Hausdorff limit

$$\lim_{n \to +\infty} (n \cdot \mathbb{M}, x_0)$$

exists, then this limit is called the tangent space to $\mathbb{M}$ at x_0.

We have the following theorem due to Mitchell [Mitchell (1985)] (see also [Pansu (1989)]):

Proposition 3.1 *Let $x_0 \in \mathbb{R}^n$ be a regular point of $\mathcal{D}$, then the tangent space at x_0 in Gromov-Hausdorff's sense exists and is equal to $\mathbf{Gr}(\mathcal{D})(x_0)$.*

Actually, even if x_0 is not a regular point of $\mathcal{D}$, the tangent space in Gromov-Hausdorff sense exists. Therefore, from this theorem, it is possible to define $\mathbf{Gr}(\mathcal{D})(x_0)$ at any point of $\mathcal{D}$. Nevertheless, if x_0 is not a regular point, then $\mathbf{Gr}(\mathcal{D})(x_0)$ is not a Lie group (see [Bellaïche (1996)]).

Example 3.12 Let us consider in $\mathbb{R}^2$, the two vector fields

$$V_1 = \frac{\partial}{\partial x},$$

and

$$V_2 = x\frac{\partial}{\partial y}.$$

These vector fields span $\mathbb{R}^2$ everywhere, except along the line $x = 0$, where adding

$$[V_1, V_2] = \frac{\partial}{\partial y}$$

is needed. So, the distribution $\mathcal{D}$ generated by V_1 and V_2 satisfies the strong Hörmander's condition. The sub-Riemannian geometry associated with $\mathcal{D}$ is called the geometry of the Grusin plane. In that case, for every $(x, y) \in \mathbb{R}^2$,

$$\mathbf{Gr}(\mathcal{D})(x, y) = \mathbb{R}^2, \text{ if } x \neq 0,$$

whereas,

$$\mathbf{Gr}(\mathcal{D})(x, y) = \mathbb{G}_2(\mathbb{R}^2)/\exp(\mathbb{R}V_2), \text{ if } x = 0.$$

Before we turn to applications in the theory of hypoelliptic diffusions, we still have to generalize the notion of Pansu's derivative (see Chapter 2, Section 2.3.) in our context of general sub-Riemannian manifolds. This generalization is due to [Margulis and Mostow (1995)], we shall only hint their notion of Pansu's derivative, since the rigorous construction is quite technical but rather clear from an intuitive point of view. Let $\mathbb{G}$ be a Carnot group and

$$F : \mathbb{G} \to \mathbb{R}^n$$

be a map. Then F is said to be Pansu differentiable at $g \in \mathbb{G}$ if there exists a Carnot group morphism

$$d_P F(g) : \mathbb{G} \to \mathbf{Gr}(\mathcal{D})(F(g)),$$

such that for small t

$$F(g\Delta_t^{\mathbb{G}}h) \approx F(g)\left(\Delta_t^{\mathbf{Gr}(\mathcal{D})(F(g))}d_P F(g)(h)\right), \quad h \in \mathbb{G},$$

where $\Delta^{\mathbb{G}}$ (resp. $\Delta^{\mathbf{Gr}(\mathcal{D})(F(g))}$) are the canonical dilations on the Carnot group $\mathbb{G}$ (resp. $\mathbf{Gr}(\mathcal{D})(F(g))$).

This approximation formula should be understood in the same way than the usual approximation formula

$$F(x + h) \approx F(x) + dF(x)(h),$$

for a map F defined between two manifolds.

We now apply all these new notions to study the stochastic differential equation

$$X_t^{x_0} = x_0 + \sum_{i=1}^d \int_0^t V_i(X_s^{x_0}) \circ dB_s^i. \tag{3.4}$$

Theorem 3.7 *Let $x \in \mathbb{R}^n$ be a regular point of $\mathcal{D}$. Let $(X_t^x)_{t\geq 0}$ denote the solution of (3.4) with initial condition x. There exists a map*

$$F : \mathbb{G}_{r(x)}(\mathbb{R}^d) \to \mathbb{R}^n$$

such that

$$X_t^x = F(B_t^*) + t^{\frac{r(x)+1}{2}}\mathbf{R}(t),$$

where:

(1) $(B_t^*)_{t\geq 0}$ is the lift of $(B_t)_{t\geq 0}$ in the group $\mathbb{G}_N(\mathbb{R}^d)$;
(2) the remainder term $\mathbf{R}(t)$ is bounded in probability when $t \to 0$.

Moreover the map F is Pansu differentiable at $0_{\mathbb{G}_{r(x)}}$ and the Pansu's derivative

$$d_P F(0_{\mathbb{G}_{r(x)}}) : \mathbb{G}_{r(x)}(\mathbb{R}^d) \to \mathbf{Gr}(\mathcal{D})(x)$$

is a surjective Carnot group morphism.

Proof.   From Theorem 2.6 of Chapter 2, we have

$$X_t^x = F(B_t^*) + t^{\frac{r(x)+1}{2}}\mathbf{R}(t), \ t \geq 0,$$

where the remainder term $\mathbf{R}$ is bounded in probability when $t \to 0$, and

$$F(B_t^*) = \left[\exp \left(\sum_{k=1}^{r(x)} \sum_{I=(i_1,\ldots,i_k)} \Lambda_I(B)_t V_I \right) \right](x).$$

Let now $g \in \mathbb{G}$ and $y : [0,1] \to \mathbb{R}^d$ an absolutely continuous path whose lift $y^* : [0,1] \to \mathbb{G}_{r(x)}(\mathbb{R}^d)$ satisfies

$$y_0^* = 0_{\mathbb{G}_{r(x)}(\mathbb{R}^d)}$$

and

$$y_1^* = g.$$

We have

$$F\left(\Delta_t^{\mathbb{G}_{r(x)}(\mathbb{R}^d)} g \right) = \left[\exp \left(\sum_{k=1}^{r(x)} t^k \sum_{I=(i_1,\ldots,i_k)} \Lambda_I(y)_1 V_I \right) \right](x).$$

Therefore, in small times

$$"F\left(\Delta_t^{\mathbb{G}_{r(x)}(\mathbb{R}^d)} g \right) \approx x + \sum_{k=1}^{r(x)} t^k \sum_{I=(i_1,\ldots,i_k)} \Lambda_I(y)_1 V_I(x)",$$

which leads to the expected result. $\qquad\qquad\qquad\qquad\qquad\square$

We have a stronger approximation result in the normal case.

Theorem 3.8 *Let $x \in \mathbb{R}^n$ be a normal point of $\mathcal{D}$. Let $(X_t^x)_{t \geq 0}$ denote the solution of (3.4) with initial condition x. There exist a surjective Carnot group morphism*

$$\pi_x : \mathbb{G}_{r(x)}(\mathbb{R}^d) \to \mathbf{Gr}(\mathcal{D})(x)$$

and a local diffeomorphism

$$\psi_x : U \subset \mathbf{Gr}(\mathcal{D})(x) \to \mathbb{R}^n$$

such that

$$X_t^x = \psi_x\left(\pi_x B_t^* \right) + t^{\frac{r(x)+1}{2}} \mathbf{R}(t), \ 0 < t < T$$

where:

(1) U is an open neighborhood of the identity element of $\mathbf{Gr}(\mathcal{D})(x)$;
(2) B^* is the lift of B in the free Carnot group $\mathbb{G}_{r(x)}(\mathbb{R}^d)$;

(3) T is an almost surely positive stopping time;
(4) $\mathbf{R}$ is bounded in probability when $t \to 0$.

Proof.   From Theorem 2.6 of Chapter 2, we have

$$X_t^x = \left[\exp \left(\sum_{k=1}^{r(x)} \sum_{I=(i_1,\ldots,i_k)} \Lambda_I(B)_t V_I \right) \right](x) + t^{\frac{r(x)+1}{2}} \mathbf{R}(t), \ t \geq 0,$$

where the remainder term $\mathbf{R}$ is bounded in probability when $t \to 0$. Now, since x is a normal point of $\mathcal{D}$, we can write

$$\left[\exp \left(\sum_{k=1}^{r(x)} \sum_{I=(i_1,\ldots,i_k)} \Lambda_I(B)_t V_I \right) \right](x) = \psi_x(\hat{B}_t), \ t < T,$$

where

(1) $\psi_x : U \subset \mathbf{Gr}(\mathcal{D})(x) \to \mathbb{R}^n$ is a local diffeomorphism;
(2) $\hat{B}$ is the lift of B in the Carnot group $\mathbf{Gr}(\mathcal{D})(x)$, with respect to the family $(V_1(x), \ldots, V_d(x))$ (recall that by construction, this family Lie generates $\mathcal{N}(\mathcal{D})(x)$) ;
(3) T is an almost surely positive stopping time.

Now, since $\mathfrak{g}_{r(x)}(\mathbb{R}^d)$ is free, there exists a unique Lie algebra surjective homomorphism $\alpha_x : \mathfrak{g}_{r(x)}(\mathbb{R}^d) \to \mathcal{N}(\mathcal{D})(x)$ such that $\alpha_x(D_i) = V_i(x)$. Since Carnot groups are simply connected nilpotent groups for which the exponential map is a diffeomorphism, there exists a unique Carnot group morphism $\pi_x : \mathbb{G}_{r(x)} \to \mathbf{Gr}(\mathcal{D})(x)$ such that $d\pi_x = \alpha_x$. We have $\pi_x(B_t^*) = \hat{B}_t$ which concludes the proof.   □

Remark 3.9   *Observe that $\pi_x B_t^*$ is a lift of B in the Carnot group $\mathbf{Gr}(\mathcal{D})(x)$.*

Remark 3.10   *Observe that the map $\psi_x : U \subset \mathbf{Gr}(\mathcal{D})(x) \to \mathbb{R}^n$ is bi-Lipschitz with respect to the respective sub-Riemannian distances of $\mathbf{Gr}(\mathcal{D})(x)$ and $\mathbb{R}^n$.*

Remark 3.11   *We stress that theorem (3.8) is not true in general if x is not a normal point of $\mathcal{D}$. Indeed, let us assume that the nilpotentization $\mathcal{N}(\mathcal{D})(x)$ is not constant in a neighborhood of x and that there exists a bi-Lipschitz map $\psi_x : U \subset \mathbf{Gr}(\mathcal{D})(x) \to \mathbb{R}^n$. In that case an extension of Pansu-Rademacher's theorem (see previous chapter) due to [Margulis and Mostow (1995)] would imply that ψ_x is almost everywhere Pansu differentiable and the derivatives would provide Carnot group morphisms between* *groups which are not isomorphic. This argument actually shows that there is no inversion local theorem with respect to the Pansu's derivative.*

A case of particular importance which is covered by the previous theorem is the elliptic case which corresponds to the case $d = n$ and $(V_1(x), ..., V_n(x))$ is a basis of $\mathbb{R}^n$ for any $x \in \mathbb{R}^n$. In that case, recall that every point is normal and we obtain

Corollary 3.1 *Let $x \in \mathbb{R}^n$ and let $(X_t^x)_{t \geq 0}$ denote the solution of (3.4) with initial condition x. There exists a local diffeomorphism*

$$\psi_x : U \subset \mathbb{R}^n \to \mathbb{R}^n$$

such that

$$X_t^x = \psi_x(B_t) + t\mathbf{R}(t), \ 0 < t < T$$

where:

(1) U is an open neighborhood of 0;
(2) T is an almost surely non negative stopping time;
(3) $\mathbf{R}$ is bounded in probability when $t \to 0$.

Another immediate corollary of Theorem 3.8 is the behaviour in small times of a hypoelliptic heat kernel on the diagonal (see [Ben Arous (1989a)], [Léandre (1992)] and [Takanobu (1988)]).

Corollary 3.2 *Let x be a regular point of $\mathcal{D}$. Let p_t, $t > 0$, denote the density of X^x with respect to the Lebesgue measure. We have,*

$$p_t(x) \sim_{t \to 0} \frac{C(x)}{t^{\frac{D(x)}{2}}},$$

where $C(x)$ is a non negative constant and $D(x)$ the Hausdorff dimension of the tangent space $\mathbf{Gr}(\mathcal{D})(x)$.

Proof. We use Theorem 3.7 to write

$$X_t^x = F(B_t^*) + t^{\frac{r(x)+1}{2}} \mathbf{R}(t),$$

where:

(1) $(B_t^*)_{t \geq 0}$ is the lift of $(B_t)_{t \geq 0}$ in the group $\mathbb{G}_N(\mathbb{R}^d)$;
(2) $\mathbf{R}(t)$ is bounded in probability when $t \to 0$;

(3) the map

$$F : \mathbb{R}^n \times \mathbb{G}_{r(x)}(\mathbb{R}^d) \to \mathbb{R}^n$$

is Pansu differentiable at the origin with a surjective Pansu's derivative.

Now, we have

$$(F(B_t^*))_{t \geq 0} =^{law} \left(F \left(\Delta_{\sqrt{t}}^{\mathbb{G}_{r(x)}(\mathbb{R}^d)} B_1^* \right) \right)_{t \geq 0},$$

and

$$F \left(\Delta_{\sqrt{t}}^{\mathbb{G}_{r(x)}(\mathbb{R}^d)} B_1^* \right) \approx x \Delta_{\sqrt{t}}^{\mathbf{Gr}(\mathcal{D})(x)} d_P F(0_{\mathbb{G}_{r(x)}(\mathbb{R}^d)})(B_1^*),$$

in small times. Since $d_P F(0_{\mathbb{G}_{r(x)}(\mathbb{R}^d)})$ is surjective, the random variable

$$d_P F(0_{\mathbb{G}_{r(x)}(\mathbb{R}^d)})(B_1^*)$$

admits a density with respect to the Lebesgue measure. This concludes the proof.   □

3.4   Horizontal diffusions

In this section, we provide interesting examples of hypoelliptic diffusions which arise naturally in Riemannian geometry. We shall assume basic knowledge in Riemannian geometry and stochastic processes in this setting as can be found in the Appendices A and B. We shall also assume some very basic knowledge of the theory of principal bundles; for further details, we refer to the Chapter 2 of [Kobayashi and Nomizu (1996)].

Let $(\mathbb{M}, g)$ be a d dimensional connected compact smooth Riemannian manifold. That is, the manifold $\mathbb{M}$ is endowed with a $(0, 2)$ definite positive tensor g. This non-degenerate metric tensor can e.g. stem from an elliptic differential system $(V_1, ..., V_d)$. That is, $V_1, ..., V_d$ are smooth vector fields on $\mathbb{M}$ such that for every $x_0 \in \mathbb{M}$, $(V_1(x_0), ..., V_d(x_0))$ is a basis of the tangent space to $\mathbb{M}$ at x_0. In that case g is defined at x_0 by the condition that $(V_1(x_0), ..., V_d(x_0))$ is an orthonormal basis.

The tangent bundle to $\mathbb{M}$ is denoted $T\mathbb{M}$ and $T_m\mathbb{M}$ is the tangent space at $m \in \mathbb{M}$: we have hence $T\mathbb{M} = \cup_m T_m\mathbb{M}$.

An orthonormal frame at $m \in \mathbb{M}$ is a linear isomorphism $u : \mathbb{R}^d \to T_m\mathbb{M}$ such that for every $x, y \in \mathbb{R}^d$,

$$g_m(u(x), u(y)) = \langle x, y \rangle_{\mathbb{R}^d},$$

where $\langle x, y \rangle_{\mathbb{R}^d}$ denotes the usual scalar product in $\mathbb{R}^d$. The set of orthonormal frames at m shall be denoted $\mathcal{O}(\mathbb{M})_m$. Observe that the Lie group $\mathcal{O}_d(\mathbb{R})$ of $d \times d$ orthogonal matrices acts naturally on $\mathcal{O}(\mathbb{M})_m$ by $u \to u \circ g$, $u \in \mathcal{O}(\mathbb{M})_m$, $g \in \mathcal{O}_d(\mathbb{R})$. Set now

$$\mathcal{O}(\mathbb{M}) = \bigcup_{m \in \mathbb{M}} \mathcal{O}(\mathbb{M})_m.$$

This set is called the orthonormal frame bundle of $\mathbb{M}$. It is easily checked that it can be endowed with a differentiable manifold structure with dimension $\frac{d(d+1)}{2}$ that makes the canonical projection $\pi : \mathcal{O}(\mathbb{M}) \to \mathbb{M}$ a smooth map. We summarize the above properties by saying that $(\mathcal{O}(\mathbb{M}), \mathbb{M}, \mathcal{O}_d(\mathbb{R}))$ is a principal bundle on $\mathbb{M}$ with structure group the group $\mathcal{O}_d(\mathbb{R})$. Working in this bundle aims at making equivariant the objects we are dealing with.

As in Appendix B, it is possible to show the existence of a unique metric, torsion free, connection on $\mathbb{M}$. That is, if we denote by $\mathcal{V}(\mathbb{M})$ the set of smooth vector fields on $\mathbb{M}$, there is a unique map

$$\nabla : \mathcal{V}(\mathbb{M}) \times \mathcal{V}(\mathbb{M}) \to \mathcal{V}(\mathbb{M})$$

such that for any $U, V, W \in \mathcal{V}(\mathbb{M})$ and any smooth $f, g : \mathbb{M} \to \mathbb{R}$:

(1) $\nabla_{fU+gV} W = f \nabla_U W + g \nabla_V W$;
(2) $\nabla_U(V + W) = \nabla_U V + \nabla_U W$;
(3) $\nabla_U(fV) = f \nabla_U V + U(f)V$;
(4) $\nabla_U V - \nabla_V U = [U, V]$;
(5) $U(g(V, W)) = g(\nabla_U V, W) + g(V, \nabla_U W)$.

A vector field X along a smooth curve $(c_t)_{t \geq 0}$ on $\mathbb{M}$ is said to be parallel along the curve if we always have

$$\nabla_{c'} X = 0.$$

A smooth curve $(u_t)_{t \geq 0}$ in $\mathcal{O}(\mathbb{M})$ is called horizontal if for every $x \in \mathbb{R}^d$, the vector field $u(x)$ is parallel along the curve πu. A vector $X \in T_{u_0}\mathcal{O}(\mathbb{M})$ is called horizontal if u is horizontal. The space of horizontal vectors at

u_0 is denoted $\mathcal{H}_{u_0}\mathcal{O}(\mathbb{M})$; it is a space of dimension d. We have the direct decomposition

$$T_{u_0}\mathcal{O}(\mathbb{M}) = \mathcal{H}_{u_0}\mathcal{O}(\mathbb{M}) \oplus \mathcal{V}_{u_0}\mathcal{O}(\mathbb{M})$$

where $\mathcal{V}_{u_0}\mathcal{O}(\mathbb{M})$ denotes the space of vectors tangent to the fiber $\mathcal{O}(\mathbb{M})_{\pi u_0}$.

The canonical projection π induces an isomorphism

$$\pi_* : \mathcal{H}_{u_0}\mathcal{O}(\mathbb{M}) \to T_{\pi u_0}\mathbb{M},$$

and for each $X \in T_m\mathbb{M}$ and a frame u_0 at m, there is a unique horizontal vector X^*, the horizontal lift of X from u_0 such that $\pi_* X^* = X$.

For each $x \in \mathbb{R}^d$ we can define a horizontal vector field H_x by the property that at each point $u \in \mathcal{O}(\mathbb{M})$, $H_x(u)$ is the horizontal lift of $u(x)$ from u.

Definition 3.8   Let $(e_1, ..., e_d)$ be the canonical basis of $\mathbb{R}^d$. The fundamental horizontal vector fields are given by

$$H_i = H_{e_i}.$$

Observe that the differential system on $\mathcal{O}(\mathbb{M})$ generated by the horizontal vector fields $H_1, ..., H_d$, is nothing else than the horizontal distribution for the Levi-Civita connection on $\mathbb{M}$, that is $\mathcal{HO}(\mathbb{M})$.

The Bochner's horizontal Laplacian is by definition the sum of squares operator given by

$$\Delta_{\mathcal{O}(\mathbb{M})} = \sum_{i=1}^d H_i^2.$$

The fundamental property of the Bochner's horizontal Laplacian is that it is the lift of the Laplacian operator $\Delta_\mathbb{M}$ of $\mathbb{M}$. That is, for every smooth $f : \mathbb{M} \to \mathbb{R}$,

$$\Delta_{\mathcal{O}(\mathbb{M})}(f \circ \pi) = (\Delta_\mathbb{M} f) \circ \pi.$$

As a direct consequence of this, it follows that the solution of the stochastic differential equation

$$B_t^* = U_0 + \sum_{i=1}^d \int_0^t H_i(B_s^*) \circ d\widetilde{B}_s^i, \quad t \geq 0,$$

where $U_0 \in \mathcal{O}(\mathbb{M})$ and $\left(\widetilde{B}_t\right)_{t \geq 0}$ is a $\mathbb{R}^d-$ valued standard Brownian motion, is such that the $\mathbb{M}$-valued process $(\pi(B_t^*))_{t \geq 0}$ is a Brownian motion. That is,

$(\pi(B_t^*))_{t\geq 0}$ is a Markov process with generator $\frac{1}{2}\Delta_{\mathbb{M}}$. The process $(B_t^*)_{t\geq 0}$ is called a horizontal Brownian motion.

Conversely, it can be shown that any Brownian motion $(B_t)_{t\geq 0}$ on $\mathbb{M}$ can be written $(\pi(B_t^*))_{t\geq 0}$ where $(B_t^*)_{t\geq 0}$ is a horizontal Brownian motion. The process $(B_t^*)_{t\geq 0}$ is then called the horizontal lift of $(B_t)_{t\geq 0}$. The linear Brownian motion $\left(\widetilde{B}_t\right)_{t\geq 0}$ which drives the stochastic differential equation driven by $(B_t^*)_{t\geq 0}$ is called the anti-development of $(B_t)_{t\geq 0}$.

This passage through the orthonormal bundle is the classical Eels-Elworthy-Malliavin's approach to Riemannian Brownian motions (see e.g. [Emery (1989)], [Hsu (2002)], or Part V of [Malliavin (1997)]). The advantage of this construction of Riemannian Brownian motions is that, firstly it is intrinsic, and secondly it provides a pathwise construction obtained by solving a globally defined stochastic differential equation. The bijective map

$$\left(\widetilde{B}_t\right)_{t\geq 0} \to (B_t)_{t\geq 0}$$

is called the Itô's map of the manifold. Also observe that this construction proves that the natural filtration of a Riemannian Brownian motion is the same than the natural filtration of a linear d-dimensional Brownian motion; a fact, which a priori was not obvious from the original definition of the Brownian motions on a manifold (see Appendix A).

Let us now see how the Levi-Civita connection ∇ manifests itself on the bundle $(\mathcal{O}(\mathbb{M}), \mathbb{M}, \mathcal{O}_d(\mathbb{R}))$. Let us denote $\mathfrak{o}_d(\mathbb{R})$ the Lie algebra of $\mathcal{O}_d(\mathbb{R})$, that is the space of $d \times d$ skew symmetric matrices. For every $M \in \mathfrak{o}_d(\mathbb{R})$, we can define a vertical vector field V_M on $\mathcal{O}(\mathbb{M})$ by

$$(V_M F)(u) = \lim_{t\to 0} \frac{F\left(ue^{tM}\right) - F(u)}{t},$$

with $u \in \mathcal{O}(\mathbb{M})$ and $F : \mathcal{O}(\mathbb{M}) \to \mathbb{R}$. Observe that the map $M \to V_M(u)$ is an isomorphism from $\mathfrak{o}_d(\mathbb{R})$ onto the set $\mathcal{V}_u\mathcal{O}(\mathbb{M})$ of vertical vector fields at u. Then, the connection form on $\mathcal{O}(\mathbb{M})$ (often called the Ehresmann connection), is the unique skew-symmetric matrix ω of one forms on $\mathcal{O}(\mathbb{M})$ such that:

(1) $\omega(X) = 0$ if and only if $X \in \mathcal{HO}(\mathbb{M})$;
(2) $V_{\omega(X)} = X$ if and only if $X \in \mathcal{VO}(\mathbb{M})$.

Associated with this Ehresmann connection on $\mathcal{O}(\mathbb{M})$, which corresponds to the Levi-Civita connection on $\mathbb{M}$ we have two Cartan's structural equations (see Appendix B, Section B.8 for the two Cartan's structural equations on M). The first structural equation reads

$$\mathbf{d}\Theta + \Theta \wedge \omega = 0,$$

where Θ is the so-called tautological one form on $\mathcal{O}(\mathbb{M})$ defined by:

$$\Theta(X)(u) = u^{-1}\pi_*X, \quad X \in T_u\mathcal{O}(\mathbb{M}).$$

The second structural equation reads

$$\mathbf{d}\omega + \omega \wedge \omega = \Omega,$$

and can be taken as a definition of the curvature form Ω. It can be shown that

$$\Omega(X,Y)(u) = u^{-1}R(\pi_*X, \pi_*Y)u, \quad X, Y \in T_u\mathcal{O}(\mathbb{M}),$$

where R denotes the Riemannian curvature tensor on $\mathbb{M}$.

For $u \in \mathcal{O}(\mathbb{M})$, let us now consider the map

$$\Psi_u : \mathfrak{o}_d(\mathbb{R}) \rightarrow \mathfrak{o}_d(\mathbb{R})$$

such that

$$\Psi_u(M) = \left(\sum_{1 \leq i < j \leq d} \Omega_l^k(H_i, H_j)(u)M_{i,j} \right)_{1 \leq k, l \leq d}.$$

We have the following theorem (compare to Chernyakov's theorem, see [Gershkovich and Vershik (1994)] pp. 22):

Theorem 3.9 *Assume that for every $u \in \mathcal{O}(\mathbb{M})$, the application Ψ_u is an isomorphism, then:*

(1) the horizontal distribution $\mathcal{HO}(\mathbb{M})$ satisfies the strong Hörmander's condition;

(2) every $u \in \mathcal{O}(\mathbb{M})$ is normal and $\mathbf{Gr}(\mathcal{HO}(\mathbb{M}))(u) = \mathbb{G}_2(\mathbb{R}^d)$.

Proof. From the Cartan's formula, we have

$$\Theta([H_i, H_j]) = H_i\Theta(H_j) - H_j\Theta(H_i) - d\Theta(H_i, H_j),$$

where Θ is the tautological one-form on $\mathcal{O}(\mathbb{M})$. Now, from the first structural equation, we have

$$\mathbf{d}\Theta = -\Theta \wedge \omega$$

where ω is the Ehresmann connection form. This implies,

$$\Theta([H_i, H_j]) = 0.$$

Thus, the commutator of two fundamental vector fields is vertical. Using the second structure equation

$$\mathbf{d}\omega = -\omega \wedge \omega + \Omega,$$

where Ω is the curvature form, we obtain in a similar way with Cartan's formula,

$$\omega([H_i, H_j]) = -\Omega(H_i, H_j).$$

Thus,

$$[H_i, H_j] = -V_{\Omega(H_i, H_j)}.$$

Since the map $M \to V_M(u)$ is an isomorphism, for every $u \in \mathcal{O}(\mathbb{M})$ the family $(H_i(u), [H_j, H_k](u))$ is a basis of $T_u\mathcal{O}(\mathbb{M})$. Therefore the horizontal distribution $\mathcal{HO}(\mathbb{M})$ satisfies the strong Hörmander's condition. The fact every $u \in \mathcal{O}(\mathbb{M})$ is normal and $\mathbf{Gr}(\mathcal{HO}(\mathbb{M}))(u) = \mathbb{G}_2(\mathbb{R}^d)$ stems from the fact the growth vector is maximal and constant on $\mathcal{O}(\mathbb{M})$. The above computations have indeed shown that it is always $\left(d, \frac{d(d-1)}{2}\right)$. $\square$

Remark 3.12 *In dimension $d = 2$, the assumption that for every u, the application Ψ_u is non degenerate is equivalent to the fact that the Gauss curvature of $\mathbb{M}$ never vanishes.*

Remark 3.13 *If we assume that $(\mathbb{M}, g)$ is an oriented manifold the assumption that at $u \in \mathcal{O}(\mathbb{M})$, the application Ψ_u is an isomorphism is equivalent to the fact that the holonomy group at $\pi(u)$ is $\mathbf{SO}_d(\mathbb{R})$.*

By applying Theorem 3.8 and using the identification $\mathbb{G}_2(\mathbb{R}^d) \simeq \mathbb{R}^d \times \mathbb{R}^{\frac{d(d-1)}{2}}$, we obtain therefore

Corollary 3.3 *Assume that for every $u \in \mathcal{O}(\mathbb{M})$, the application Ψ_u is an isomorphism. Let $(B_t)_{t\geq 0}$ be a Brownian motion on $\mathbb{M}$. Then, there exist a local submersion*

$$\psi : U \subset \mathbb{R}^d \times \mathbb{R}^{\frac{d(d-1)}{2}} \to \mathcal{O}(\mathbb{M})$$

and a standard linear Brownian motion $(\beta_t)_{t\geq0}$ on $\mathbb{R}^d$, such that for any smooth function $f : \mathbb{M} \to \mathbb{R}$,

$$f(B_t) = f\left(\psi\left(\beta_t, \left(\int_0^t \beta_s^i d\beta_s^j - \beta_s^j d\beta_s^i\right)_{1\leq i<j\leq d}\right)\right) + t^{\frac{3}{2}}\mathbf{R}(t,f), \quad 0 < t < T,$$

where:

(1) U is an open neighborhood of 0 in $\mathbb{R}^d \times \mathbb{R}^{\frac{d(d-1)}{2}}$;
(2) T is an almost surely non negative stopping time;
(3) $\mathbf{R}$ is bounded in probability when $t \to 0$.

Proof.   Indeed, let $(B_t)_{t\geq0}$ be a Brownian motion on $\mathbb{M}$. Let us denote $(B_t^*)_{t\geq0}$ the horizontal lift of $(B_t)_{t\geq0}$ to $\mathcal{O}(\mathbb{M})$ and $(\beta_t)_{t\geq0}$ the anti-development of $(B_t)_{t\geq0}$. By applying Theorem 3.8 to $(B_t^*)_{t\geq0}$, we obtain the expected result.   □

Under the non-degeneracy assumption, we can moreover give an expression for the horizontal heat kernel in small times. On $\mathcal{O}(\mathbb{M})$, let us consider the measure μ which reads in local bundle trivializations as the product of the Riemannian measure on $\mathbb{M}$ and of the normalized Haar measure of $\mathcal{O}_d(\mathbb{R})$.

Corollary 3.4   *Assume that for every $u \in \mathcal{O}(\mathbb{M})$, the application Ψ_u is an isomorphism. Then, if p_t^* denotes the density with respect to μ of a horizontal Brownian motion on $\mathcal{O}(\mathbb{M})$ started at u, we have*

$$p_t(u) \simeq_{t\to0} \frac{C}{|\det \Psi_u| \, t^{\frac{d^2}{2}}},$$

where $C > 0$ is a universal constant which does not depend on $\mathbb{M}$.

Proof.   It follows readily from the approximation

$$f(B_t^*) = f\left(\exp\left(\sum_{i=1}^d H_i\beta_t^i + \frac{1}{2}\sum_{1\leq j<k\leq d}[H_j, H_k]\int_0^t \beta_s^i d\beta_s^j - \beta_s^j d\beta_s^i\right)(u)\right)$$
$$+ t^{\frac{3}{2}}\mathbf{R}(t),$$

which stems from Theorem 3.8 (here $f : \mathcal{O}(\mathbb{M}) \to \mathbb{R}$ is a smooth function and β the anti-development of B).   □

We can extend without difficulties the previous discussion to the case of general principal bundles. Let $\mathbb{M}$ be a d-dimensional connected smooth manifold. Let now $\mathbb{G}$ be a Lie subgroup of the linear group $\mathbf{GL}_d(\mathbb{R})$ with Lie algebra $\mathfrak{g}$. A principal bundle $(\mathbb{B}, \mathbb{M}, \mathbb{G})$ consists first of a submersion

$$\pi : \mathbb{B} \to \mathbb{M}$$

such that each fiber is isomorphic to $\mathbb{G}$. There is also a right action of $\mathbb{G}$ on $\mathbb{B}$ with the property that orbits are exactly the fibers of the bundle. Finally, we must ask that $(\mathbb{B}, \mathbb{M}, \mathbb{G})$ is locally trivializable in the sense that for every $x \in \mathbb{M}$, there is a diagram

$$\pi^{-1}(U) \xrightarrow{\phi} U \times G$$
$$\searrow \qquad \swarrow$$
$$U$$

that commutes with the right actions of $\mathbb{G}$ on both $\pi^{-1}(U)$ and $U \times G$, where ϕ is a local diffeomorphism and U a neighborhood x. For $u_0 \in \mathbb{B}$, we denote $\mathcal{V}_{u_0}\mathbb{B}$ the space of vectors tangent to the fiber $\mathbb{B}_{\pi u_0}$. As before, for every $X \in \mathfrak{g}$, we can define a vertical vector field V_X on $\mathbb{B}$ by

$$(V_X F)(u) = \lim_{t \to 0} \frac{F\left(ue^{tX}\right) - F(u)}{t}, \quad u \in \mathbb{B}, \quad F : \mathbb{B} \to \mathbb{R},$$

and the map $X \to V_X(u)$ is an isomorphism from $\mathfrak{g}$ onto the set $\mathcal{V}_u\mathcal{O}(\mathbb{M})$.

A connection in $(\mathbb{B}, \mathbb{M}, \mathbb{G})$ is an assignment of a subspace $\mathcal{H}_{u_0}\mathbb{B}$ of $\mathrm{T}_{u_0}\mathbb{B}$ to each $u_0 \in \mathbb{B}$ such that:

(1) $\mathrm{T}_{u_0}\mathbb{B} = \mathcal{H}_{u_0}\mathbb{B} \oplus \mathcal{V}_{u_0}\mathbb{B}$;
(2) $\mathcal{H}_{u_0 g}\mathbb{B} = (\mathbf{R}_g)_* \mathcal{H}_{u_0}\mathbb{B}$, for every $u_0 \in \mathbb{B}$ and $g \in \mathbb{G}$, where $\mathbf{R}_g$ is the transformation of $\mathbb{B}$ induced by $g \in \mathbb{G}$, $\mathbf{R}_g u_0 = u_0 g$;
(3) $\mathcal{H}_{u_0 g}\mathbb{B}$ depends differentiably on u_0.

Given a connection in $(\mathbb{B}, \mathbb{M}, \mathbb{G})$ we define the connection form ω as the $\mathfrak{g}$ valued one-form on $\mathbb{B}$ such that:

(1) $\omega(X) = 0$ if and only if $X \in \mathcal{H}\mathbb{B}$;
(2) $V_{\omega(X)} = X$ if and only if $X \in \mathcal{V}\mathbb{B}$.

We also define the curvature form Ω as the $\mathfrak{g}$ valued two-form on $\mathbb{B}$ such that:

$$\mathbf{d}\omega + \omega \wedge \omega = \Omega.$$

As in the orthonormal frame bundle over a Riemannian manifold, we can define d fundamental horizontal vector fields $H_1, ..., H_d$. Indeed, for every $u \in \mathbb{B}$, the canonical projection π induces an isomorphism

$$\pi_* : \mathcal{H}_u \mathbb{B} \to \mathrm{T}_{\pi u} \mathbb{M},$$

and for each $X \in \mathrm{T}_m \mathbb{M}$ and $u_0 \in \mathbb{B}$ at m, there is a unique horizontal vector X^*, the horizontal lift of X from u_0 such that $\pi_* X^* = X$. Now, for each $x \in \mathbb{R}^d$ we can define a horizontal vector field H_x by the property that at each point $u \in \mathbb{B}$, $H_x(u)$ is the horizontal lift of $u(x)$ from u. The fundamental horizontal vector fields are then given by $H_i = H_{e_i}$, where $(e_1, ..., e_d)$ is the canonical basis of $\mathbb{R}^d$. Observe that the differential system on $\mathbb{B}$ generated by the horizontal vector fields $H_1, ..., H_d$ is $\mathcal{H}\mathbb{B}$.

We define finally the fundamental horizontal diffusion on the bundle $(\mathbb{B}, \mathbb{M}, \mathbb{G})$ as the diffusion process on $\mathbb{B}$ with infinitesimal generator

$$\Delta_{\mathbb{B}} = \sum_{i=1}^{d} H_i^2. \tag{3.5}$$

The hypoellipticity property of $\Delta_{\mathbb{B}}$ is again related to the non-degenerence of the curvature form Ω. Namely, for $u \in \mathbb{B}$, consider the map

$$\Psi_u : \mathfrak{o}_d(\mathbb{R}) \to \mathfrak{g}$$

defined by

$$\Psi_u(M) = \sum_{1 \leq i < j \leq d} M_{i,j} \Omega(H_i, H_j)(u).$$

Proposition 3.2 *Let us assume that at $u \in \mathbb{B}$, Ψ_u is surjective, then the operator $\Delta_{\mathbb{B}}$ satisfies the Hörmander's condition at u. More precisely, $(H_i(u), [H_j, H_k](u))_{1 \leq i,j,k \leq d}$ generates $\mathrm{T}_u \mathbb{B}$.*

Proof. It directly stems from the identity

$$[H_i, H_j] = -V_{\Omega(H_i, H_j)}.$$

$\square$

If Ψ_u is always surjective then all the points are normal and we can moreover compute explicitly the nilpotentizations and the tangent spaces. Let $u \in \mathbb{B}$.

Consider on the linear space $\mathbb{R}^d \oplus \mathfrak{g}$ the polynomial group law given by

$$(x, X) \star (y, Y) = \left(x + y, X + Y + \frac{1}{2} \Omega \left(\sum_{i=1}^{d} x_i H_i, \sum_{i=1}^{d} y_i H_i \right)(u) \right).$$

Then, it is easily seen that $(\mathbb{R}^d \oplus \mathfrak{g}, \star)$ is a two-step Carnot group that we shall denote $\mathbb{G}\Omega_u$. Observe that from the assumption that Ψ_u is always surjective, for every $u, v \in \mathbb{B}$, there exists a Carnot group isomorphism $\mathbb{G}\Omega_u \to \mathbb{G}\Omega_v$. In this setting, we easily prove:

Proposition 3.3 *Let us assume that for every $u \in \mathbb{B}$, Ψ_u is surjective. Then every point $u \in \mathbb{B}$ is normal and $\mathbf{Gr}(\mathcal{H}\mathbb{B})(u) = \mathbb{G}\Omega_u$.*

As an immediate corollary, we get

Corollary 3.5 *Let us assume that for every $u \in \mathbb{B}$, Ψ_u is surjective. Let p_t, $t > 0$, denote the heat kernel of (3.5) with respect to any Lebesgue measure. We have,*

$$p_t(u, u) \sim_{t \to 0} \frac{C(u)}{t^{\frac{d}{2} + \dim \mathfrak{g}}},$$

where $C(u)$ is a non negative constant.

3.5   Regular sublaplacians on compact manifolds

Let $\mathbb{M}$ be a connected compact smooth manifold. We consider on $\mathbb{M}$ a second order differential operator

$$\mathcal{L} = \sum_{i=1}^{d} V_i^2,$$

which satisfies the strong Hörmander's condition. Let $\mathcal{D}$ denote the differential system generated by $V_1, ..., V_d$. We assume that there exists a Carnot group $\mathbb{G}$ such that for every $x \in \mathbb{M}$,

$$\mathbf{Gr}(\mathcal{D})(x) = \mathbb{G}.$$

If the previous assumptions are satisfied, then $\mathcal{L}$ shall be said to be regular. Since the manifold $\mathbb{M}$ is assumed to be compact, it is possible to develop a spectral theory for $\mathcal{L}$ which is similar to the spectral theory of elliptic operators. Let us hint this theory.

Let us denote X the diffusion associated with $\mathcal{L}$. First, we are going to show that X is Harris recurrent. That is, there is a Borel measure m on $\mathbb{M}$ such that for every Borel set $A \subset \mathbb{M}$:

(1) $m(A) > 0$ implies that for any $x \in \mathbb{M}$, and $t > 0$,

$$\mathbb{P}\left(\exists\, s > t,\ X_s \in A \mid X_0 = x\right) = 1;$$

(2)

$$\int_{\mathbb{M}} \mathbb{P}\left(X_t \in A \mid X_0 = x\right) m(dx) = m(A),\ t > 0.$$

For further details on Harris recurrent processes, we refer to [Revuz and Yor (1999)], pp. 425. Since $\mathbb{M}$ is compact, this recurrence property is a consequence of the following lemma:

Lemma 3.1   *Let $\mathcal{O} \subset \mathbb{M}$ be a non empty, relatively compact, connected with smooth, non empty boundary, open set, then for any $x \in \mathbb{M}$, and $t > 0$,*

$$\mathbb{P}\left(\exists\, s > t,\ X_s \in \mathcal{O} \mid X_0 = x\right) = 1.$$

Proof.   Define

$$\tau_{\mathcal{O}} = \inf\{t > 0, X_t \in \mathcal{O}\},$$

and

$$h_{\mathcal{O}}(x) = \mathbb{P}\left(\tau_{\mathcal{O}} < +\infty \mid X_0 = x\right),\ x \in \mathbb{M}.$$

It is easy to see that

$$\mathcal{L}h_{\mathcal{O}} = 0$$

on $\mathbb{M} - \mathcal{O}$. From the general theory of Markov processes, there are two possibilities: Either $h_{\mathcal{O}} = 1$ on $\mathbb{M}$ and for any $x \in \mathbb{M}$, and $t > 0$,

$$\mathbb{P}\left(\exists\, s > t,\ X_s \in \mathcal{O} \mid X_0 = x\right) = 1,$$

or $0 < h_{\mathcal{O}} < 1$ on $\mathbb{M} - \mathcal{O}$. The Bony's strong maximal principle for hypoelliptic operators (see [Bony (1969)]), implies actually that the only possibility is $h = 1$.   □

Therefore, the diffusion X is ergodic, i.e. for any smooth function $f : \mathbb{M} \to \mathbb{R}$ and any $x \in \mathbb{M}$,

$$\lim_{t \to +\infty} \mathbb{E}\left(f(X_t) \mid X_0 = x\right) = \int_{\mathbb{M}} f\, dm,$$

where m is the invariant measure for X. Observe that m is solution of the equation

$$\mathcal{L}^* m = 0,$$

so that, from Hörmander's theorem it admits a smooth density. An immediate corollary of the ergodicity of X is the following Liouville's type theorem.

Corollary 3.6 *Let $f : \mathbb{M} \to \mathbb{R}$ be a smooth function such that*

$$\mathcal{L}f = 0,$$

then f is constant.

Proof. Indeed,

$$\mathcal{L}f = 0$$

implies that for every $t \geq 0$, $x \in \mathbb{M}$,

$$\mathbb{E}\left(f(X_t) \mid X_0 = x\right) = f(x).$$

Thus,

$$f(x) = \lim_{t \to +\infty} \mathbb{E}\left(f(X_t) \mid X_0 = x\right) = \int_{\mathbb{M}} f \, dm. \qquad \square$$

We shall now assume furthermore than $\mathcal{L}$ is self-adjoint with respect to m, i.e. for any smooth functions $f, g : \mathbb{M} \to \mathbb{R}$

$$\int_{\mathbb{M}} g(\mathcal{L}f) dm = \int_{\mathbb{M}} (\mathcal{L}g) f \, dm.$$

In that case, $e^{t\mathcal{L}}$ is a compact self-adjoint operator in $\mathbf{L}^2(\mathbb{M}, m)$. We deduce that $\mathcal{L}$ has a discrete spectrum tending to $-\infty$. We furthermore have a Minakshisundaram-Pleijel type expansion for the heat kernel p_t associated with $\mathcal{L}$:

$$p_t(x, y) = \sum_{k=0}^{+\infty} e^{-\mu_k t} \left(\sum_{i=1}^{N_k} \varphi_i^k(x) \, \varphi_i^k(y) \right)$$

where:

(1) $\{-\mu_k\}$ is the set of eigenvalues of $\mathcal{L}$;
(2) $N_k = \dim \mathcal{V}(\mu_k) = \dim\{f \mid \mathcal{L}f = -\mu_k f\}$;

(3) $\left(\varphi_i^k\right)$ is an orthonormal basis of $\mathcal{V}(\mu_k)$ for the global scalar product

$$\langle \varphi, \psi \rangle = \int_M \varphi\psi dm.$$

Let us denote $\mathbf{Sp}(\mathcal{L})$ the set of eigenvalues of $\mathcal{L}$ repeated according to multiplicity.

Theorem 3.10 *For $\lambda > 0$, let*

$$\mathbf{N}(\lambda) = \mathbf{Card}\left(\mathbf{Sp}(\mathcal{L}) \cap [-\lambda, 0]\right).$$

We have

$$\mathbf{N}(\lambda) \sim_{\lambda \to +\infty} C(\mathcal{L}, \mathbf{M})\lambda^{\frac{D}{2}},$$

where $C(\mathcal{L}, \mathbf{M})$ is a non negative constant and D the Hausdorff dimension of $\mathbf{G}$.

Proof.   The asymptotic development of the heat semigroup $e^{t\mathcal{L}}$ on the diagonal leads to

$$\mathbf{Tr}(e^{t\mathcal{L}}) = \int_M p_t(x, x)dm(x) \sim_{t \to 0} \frac{K}{t^{\frac{D}{2}}},$$

where K is a non negative constant. On the other hand,

$$\mathbf{Tr}(e^{t\mathcal{L}}) = \sum_{k=0}^{+\infty} N_k e^{-\mu_k t}.$$

Therefore,

$$\sum_{k=0}^{+\infty} N_k e^{-\mu_k t} \sim_{t \to 0} \frac{K}{t^{\frac{D}{2}}}.$$

The result follows then from the following Karamata's theorem: if μ is a Borel measure on $[0, \infty)$, $\alpha \in (0, +\infty)$, then

$$\int_0^{+\infty} e^{-t\lambda} d\mu(\lambda) \sim_{t \to 0} \frac{1}{t^\alpha},$$

implies

$$\int_0^x d\mu(\lambda) \sim_{x \to +\infty} \frac{x^\alpha}{\Gamma(1+\alpha)}.$$

$\square$

Remark 3.14 *We believe that the constant $C(\mathcal{L}, \mathbb{M})$ is an interesting invariant of the sub-Riemannian geometry that $\mathcal{L}$ induces on $\mathbb{M}$ (recall that in the Riemannian case, it is simply, up to scale, the Riemannian volume of the manifold). For instance, it would be interesting to know if $C(\mathcal{L}, \mathbb{M})$ is the Hausdorff measure of $\mathbb{M}$.*

To conclude this section, this is maybe interesting to study carefully an example. Let us consider the Lie group $\mathbf{SU}(2)$. As already observed, a basis of $\mathfrak{su}(2)$ is formed by

$$V_1 = \frac{1}{2} \begin{pmatrix} i & 0 \\ 0 & -i \end{pmatrix}, \ V_2 = \frac{1}{2} \begin{pmatrix} 0 & 1 \\ -1 & -0 \end{pmatrix}, \ V_3 = \frac{1}{2} \begin{pmatrix} 0 & i \\ i & 0 \end{pmatrix},$$

and the commutation relations hold

$$[V_1, V_2] = V_3, \ [V_2, V_3] = V_1, \ [V_3, V_1] = V_2. \tag{3.6}$$

We want to study the regular sub-Laplacian

$$\mathcal{L} = V_2^2 + V_3^2.$$

Actually, we shall study the following family of operators defined for $\varepsilon \in [0, 1]$,

$$\mathcal{L}^\varepsilon = \varepsilon V_1^2 + V_2^2 + V_3^2.$$

Observe that each $\mathcal{L}^\varepsilon$ is self-adjoint with respect to the normalized Haar measure of $\mathbf{SU}(2)$. For $\varepsilon > 0$, $\mathcal{L}^\varepsilon$ is elliptic so that

$$\mathbf{Card}\,(\mathbf{Sp}(\mathcal{L}^\varepsilon) \cap [-\lambda, 0]) \sim_{\lambda \to +\infty} C_\varepsilon \lambda^{\frac{3}{2}},$$

whereas

$$\mathbf{Card}\,(\mathbf{Sp}(\mathcal{L}) \cap [-\lambda, 0]) \sim_{\lambda \to +\infty} C\lambda^2.$$

Therefore, it is interesting to understand the spectrum when $\varepsilon \to 0$.

Proposition 3.4 *Let $\varepsilon \in [0, 1)$. The set of eigenvalues of $\mathcal{L}^\varepsilon$ is the set*

$$\{-\lambda_{n,m}^\varepsilon, \ n \in \mathbb{N}, \ 0 \le m \le n\},$$

where

$$\lambda_{n,m}^\varepsilon = \varepsilon \frac{n^2}{4} + \frac{n}{2} - (1 - \varepsilon)m^2 + (1 - \varepsilon)mn.$$

Moreover, the multiplicity of $\lambda_{n,m}^\varepsilon$ is equal to $n + 1$.

Proof. We use the theory of representations of $\mathbf{SU}(2)$; for a detailed account on it, we refer to Taylor [Taylor (1986)], Chapter 2. Let $\varepsilon \in [0, 1)$. Thanks to the relations (3.6), note that $\mathcal{L}^\varepsilon$ commutes with $\mathcal{L}^1 = V_1^2 + V_2^2 + V_3^2$. Therefore $\mathcal{L}^\varepsilon$ acts on each eigenspace of $\mathcal{L}^1$. We can examine the spectrum of $\mathcal{L}^\varepsilon$ by decomposing $\mathbf{L}^2(\mathbf{SU}(2))$ into eigenspaces of $\mathcal{L}^1$, which is equivalent to decomposing it into subspaces irreducible for the regular action of $\mathbf{SU}(2) \times \mathbf{SU}(2)$ given by

$$((g, h) \cdot f)(x) = f(g^{-1}xh), \ f \in \mathbf{L}^2(\mathbf{SU}(2)), \ g, h, x \in \mathbf{SU}(2).$$

Now, it is known (see for instance Taylor [Taylor (1986)]) that, up to equivalence, for every $k \in \mathbb{N}$, there exists one and only one irreducible representation $\pi_k : \mathbf{SU}(2) \to \mathbb{C}^{k+1}$. Thus, by the Peter-Weyl theorem, the irreducible spaces of $\mathbf{L}^2(\mathbf{SU}(2))$ for the regular action are precisely the spaces of the form

$$\mathcal{V}_k = \mathbf{span}\{\pi_k^{i,j}, 1 \leq i, j \leq k+1\},$$

where $\pi_k^{i,j}$ denotes the components of the representation π^k in a chosen orthonormal basis of $\mathbb{C}^{k+1}$. Observe now that each space $\mathcal{V}_k$ is an eigenspace of $\mathcal{L}^1$. The associated eigenvalue is $-\frac{k(k+2)}{4}$. If we consider now the left regular representation

$$(g \cdot f)(x) = f(g^{-1}x), \ f \in \mathbf{L}^2(\mathbf{SU}(2)), \ g, x \in \mathbf{SU}(2),$$

then $\mathcal{V}_k$ is a direct sum of $k+1$ irreducible representations of $\mathbf{SU}(2)$, each equivalent to π_k:

$$\mathcal{V}_k = \bigoplus_{l=1}^{k+1} \mathcal{V}_{k,l},$$

where $\mathcal{V}_{k,l} = \mathbf{span}\{\pi_k^{i,l}, 1 \leq i \leq k+1\}$. Each $\mathcal{V}_{k,l}$ splits into one-dimensional eigenspaces for V_1:

$$\mathcal{V}_{k,l} = \bigoplus_{\mu} \mathcal{V}_{k,l,\mu},$$

where

$$\mu \in \left\{ -\frac{k}{2}, -\frac{k}{2} + 1, ..., \frac{k}{2} \right\},$$

and

$$V_1 = i\mu \text{ on } \mathcal{V}_{k,l,\mu}.$$

Since $\mathcal{L}^\varepsilon = \mathcal{L}^1 - (1-\varepsilon)V_1^2$, we have

$$\mathcal{L}^\varepsilon = -\left(\frac{k(k+2)}{4} - (1-\varepsilon)\mu^2\right),$$

on $\mathcal{V}_{k,l,\mu}$. $\qquad\qquad\qquad\qquad\qquad\qquad\qquad\qquad\qquad\qquad\qquad\square$

From this, we deduce immediately:

$$\mathbf{Card}\left(\mathbf{Sp}(\mathcal{L}^\varepsilon) \cap [-\lambda, 0]\right) \sim_{\lambda \to +\infty} \frac{8}{3}\sqrt{\varepsilon}\lambda^{\frac{3}{2}},$$

whereas

$$\mathbf{Card}\left(\mathbf{Sp}(\mathcal{L}) \cap [-\lambda, 0]\right) \sim_{\lambda \to +\infty} 2\lambda^2.$$

3.6 Stochastic differential equations driven by loops

On the free Carnot group $\mathbb{G}_N(\mathbb{R}^d)$, consider the fundamental process $(B_t^*)_{t\geq 0}$ defined as the solution of the stochastic differential equation

$$B_t^* = \sum_{i=1}^d \int_0^t D_i(B_s^*) \circ dB_s^i, \ t \geq 0.$$

That is, $(B_t^*)_{t\geq 0}$ is the lift in $\mathbb{G}_N(\mathbb{R}^d)$ of the Brownian motion $(B_t)_{t\geq 0}$. More generally, if $(M_t)_{t\geq 0}$ is a $\mathbb{R}^d$-valued semimartingale, we shall denote $(M_t^*)_{t\geq 0}$ the lift of $(M_t)_{t\geq 0}$ in $\mathbb{G}_N(\mathbb{R}^d)$.

Let us denote $p_t(x,y)$, $t > 0$, the smooth transition kernel with respect to the Lebesgue measure of $(B_t^*)_{t\geq 0}$. Let us recall (see Appendix A) that it is defined by the property that

$$\mathbb{P}\left(B_{t+s}^* \in dy \mid B_s^* = x\right) = p_t(x,y)dy,$$

for any $t, s > 0$ and $x, y \in \mathbb{G}_N(\mathbb{R}^d)$. From a partial differential equations point of view, it is also the fundamental solution of the equation

$$\frac{\partial p}{\partial t} = \frac{1}{2}\left(\sum_{i=1}^d D_i^2\right) p.$$

In this first proposition we construct the N-step Brownian loop. This process is the Brownian motion $(B_t)_{t\geq 0}$ conditioned by $B_T^* = 0_{\mathbb{G}_N(\mathbb{R}^d)}$.

Proposition 3.5   *Let $T > 0$. There exists a unique $\mathbb{R}^d$-valued continuous process $(P_{t,T}^N)_{0 \le t \le T}$ such that*

$$P_{t,T}^{N,i} = B_t^i + \int_0^t D_i \ln p_{T-s} \left(P_{s,T}^{N,*}, 0_{\mathbf{G}_N(\mathbb{R}^d)} \right) ds, \quad t < T, \quad i = 1, ..., d. \quad (3.7)$$

It enjoys the following properties:

(1) $P_{T,T}^{N,*} = 0_{\mathbf{G}_N(\mathbb{R}^d)}$, *almost surely;*
(2) *for any predictable and bounded functional F,*

$$\mathbb{E}\left(F\left((B_t)_{0 \le t \le T} \right) \mid B_T^* = 0_{\mathbf{G}_N(\mathbb{R}^d)} \right) = \mathbb{E}\left(F\left((P_{t,T}^N)_{0 \le t \le T} \right) \right);$$

(3) $(P_{t,T}^N)_{0 \le t \le T}$ *is a semimartingale up to time T.*

Proof.   Let us consider the Wiener space of continuous paths:

$$\left(\mathcal{C}([0,T], \mathbb{R}^d), (\mathcal{X}_t)_{0 \le t \le T}, \mathcal{X}_T, \mathbb{P} \right)$$

where:

(1) $\mathcal{C}([0,T], \mathbb{R}^d)$ is the space of continuous functions $[0,T] \to \mathbb{R}^d$;
(2) $(X_t)_{t \ge 0}$ is the coordinate process defined by $X_t(f) = f(t)$, $f \in \mathcal{C}([0,\bar{T}], \mathbb{R}^d)$;
(3) $\mathbb{P}$ is the Wiener measure on $[0,T]$, that is the law of a d-dimensional standard Brownian motion indexed by the time interval $[0,T]$;
(4) $(\mathcal{B}_t)_{0 \le t \le T}$ is the ($\mathbb{P}$-completed) natural filtration of $(B_t)_{0 \le t \le T}$.

Let $\mathbb{Q}$ denote the probability measure on $\mathcal{X}_T$ that is the law of the process $(B_t)_{t \ge 0}$ conditioned by $B_T^* = 0_{\mathbf{G}_N(\mathbb{R}^d)}$. Of course, $\mathbb{Q}$ is not equivalent to the Wiener measure $\mathbb{P}$ on $\mathcal{X}_T$, but the following equivalence relations take place

$$d\mathbb{Q}_{/\mathcal{X}_t} = \frac{p_{T-t}\left(X_t^*, 0_{\mathbf{G}_N(\mathbb{R}^d)} \right)}{p_T\left(0_{\mathbf{G}_N(\mathbb{R}^d)}, 0_{\mathbf{G}_N(\mathbb{R}^d)} \right)} d\mathbb{Q}_{/\mathcal{X}_t}, \quad t < T. \quad (3.8)$$

Therefore from Girsanov's theorem (see Appendix A), under the probability $\mathbb{Q}$, the process

$$X_t^i - \int_0^t D_i \ln p_{T-s} \left(X_s^*, 0_{\mathbf{G}_N(\mathbb{R}^d)} \right) ds, \quad t < T, \quad i = 1, ..., d,$$

is a standard Brownian motion. It means that the law of the process $(P_{t,T}^N)_{0 \le t < T}$ that solves equation (3.7) is exactly $\mathbb{Q}$. The properties (1) and (2) follow readily.

Let us now prove the semimartingale property up to time T. In the elliptic case [Bismut (1984a)] deals with the end point singularity of the Brownian bridge on a Riemannian manifold by proving an estimate of the logarithmic derivatives of the elliptic heat kernel. Unfortunately, such an estimate does not seem to be known in the hypoelliptic case, so that we have to deal by hands with the end point singularity of our equation. To prove that $(P_{t,T}^{N,*})_{0 \le t \le T}$ is a semimartingale up to time T, we need to show that for any $1 \le i \le d$,

$$\int_0^T |\, D_i \ln p_{T-s}\left(P_{s,T}^{N,*}, 0_{\mathbf{G}_N(\mathbb{R}^d)}\right) |\, ds < +\infty$$

with probability 1. To prove this, it is actually enough to check that

$$\mathbb{E}\left(\int_{\frac{T}{2}}^T |\, D_i \ln p_{T-s}\left(P_{s,T}^{N,*}, 0_{\mathbf{G}_N(\mathbb{R}^d)}\right) |\, ds\right) < +\infty.$$

Since the processes $(P_{T-t,T}^{N,*})_{0 < t \le T}$ and $(P_{t,T}^{N,*})_{0 \le t < T}$ have the same law (they have the same finite dimensional distributions), we have

$$\mathbb{E}\left(\int_{\frac{T}{2}}^T |D_i \ln p_{T-s}\left(P_{s,T}^{N,*}, 0_{\mathbf{G}_N(\mathbb{R}^d)}\right)|ds\right)$$

$$=\mathbb{E}\left(\int_0^{\frac{T}{2}} |D_i \ln p_s\left(P_{s,T}^{N,*}, 0_{\mathbf{G}_N(\mathbb{R}^d)}\right)|ds\right).$$

Using now (3.8), we get

$$\mathbb{E}\left(\int_{\frac{T}{2}}^T |D_i \ln p_{T-s}\left(P_{s,T}^{N,*}, 0_{\mathbf{G}_N(\mathbb{R}^d)}\right)|ds\right)$$

$$=\mathbb{E}\left(\int_0^{\frac{T}{2}} |\, D_i \ln p_s\left(B_s^*, 0_{\mathbf{G}_N(\mathbb{R}^d)}\right) |\, ds \frac{p_{\frac{T}{2}}\left(B_{\frac{T}{2}}^*, 0_{\mathbf{G}_N(\mathbb{R}^d)}\right)}{p_T\left(0_{\mathbf{G}_N(\mathbb{R}^d)}, 0_{\mathbf{G}_N(\mathbb{R}^d)}\right)}\right).$$

Therefore, from the Proposition 2.4 of Chapter 2, it remains then to check that

$$\mathbb{E}\left(\int_0^{\frac{T}{2}} |\, D_i \ln p_s(B_s^*, 0_{\mathbf{G}_N(\mathbb{R}^d)}) |\, ds\right) < +\infty.$$

We claim now that

$$\mathbb{E}\left(|\, D_i \ln p_s(B_s^*, 0_{\mathbf{G}_N(\mathbb{R}^d)}) |\right) \le \frac{1}{s}\mathbb{E}(|\, B_s^i \,|).$$

Indeed, an integration by parts formula proved in [Driver and Thalmaier (2001)] (see also [Elworthy and Li (1994)]) says that for any bounded measurable function f,

$$\mathbb{E}\left(f(B_s^*)(D_i \ln p_s(B_s^*, 0_{\mathbb{G}_N(\mathbb{R}^d)}))\right) = -\frac{1}{s}\mathbb{E}\left(f(B_s^*)B_s^i\right).$$

And the above inequality is proved by taking for the sign function of $D_i \ln p_s(x, 0_{\mathbb{G}_N(\mathbb{R}^d)})$.

Therefore,

$$\mathbb{E}\left(\int_0^{\frac{T}{2}} \mid D_i \ln p_s(B_s^*, 0) \mid ds\right) \leq \int_0^{\frac{T}{2}} \frac{1}{s}\mathbb{E}(\mid B_s^i \mid)ds \leq C \int_0^{\frac{T}{2}} \frac{ds}{\sqrt{s}} < +\infty,$$

which concludes the proof. $\qquad\square$

Definition 3.9 The process $(P_{t,T}^N)_{0 \leq t \leq T}$ shall be called the Brownian loop of depth N.

Example 3.13 The process $(P_{t,T}^1)_{0 \leq t \leq T}$ is simply the d-dimensional Brownian bridge from 0 to 0 with length T, that is the solution of the stochastic differential equation

$$P_{t,T}^1 = B_t - \int_0^t \frac{P_{s,T}^1}{T-s}ds.$$

Observe that this equation can be explicitly solved and leads to the following expression

$$P_{t,T}^1 = (T-t)\int_0^t \frac{dB_s}{T-s}.$$

Example 3.14 The process $(P_{t,T}^2)_{0 \leq t \leq T}$ is the d-dimensional standard Brownian motion $(B_t)_{0 \leq t \leq T}$ conditioned by

$$\left(B_T, \frac{1}{2}\left(\int_0^T B_s^i \circ dB_s^j - B_s^j \circ dB_s^i\right)_{1 \leq i,j \leq d}\right) = 0.$$

Remark 3.15 *Notice that in law,*

$$(P_{t,T}^N)_{0 \leq t \leq T} = (\sqrt{T}P_{\frac{t}{T},1}^N)_{0 \leq t \leq T}. \tag{3.9}$$

Consider now on $\mathbb{R}^n$ stochastic differential equations of the type

$$X_t = x_0 + \sum_{i=1}^d \int_0^t V_i(X_s) \circ dP_{s,T}^{i,N}, \ t \leq T \tag{3.10}$$

where:

(1) $x_0 \in \mathbb{R}^n$;
(2) $V_1, ..., V_d$ are C^∞ bounded vector fields on $\mathbb{R}^n$;
(3) $(P_{t,T}^{1,N}, ..., P_{t,T}^{d,N})_{0 \le t \le T}$ is a d -dimensional N-step Brownian loop from 0 to 0 with length $T > 0$.

As we already did, we denote by $\mathfrak{L}$ the Lie algebra generated by the vector fields V_i's and for $p \ge 2$, by $\mathfrak{L}^p$ the Lie subalgebra defined by

$$\mathfrak{L}^p = \{[X, Y], \ X \in \mathfrak{L}^{p-1}, Y \in \mathfrak{L}\}.$$

Moreover if $\mathfrak{a}$ is a subset of $\mathfrak{L}$, we denote

$$\mathfrak{a}(x) = \{V(x), V \in \mathfrak{a}\}, \ x \in \mathbb{R}^n.$$

Proposition 3.6 *For every* $x_0 \in \mathbb{R}^n$, *there is a unique solution* $(X_t^{x_0})_{0 \le t \le T}$ *to (3.10). Moreover there exists a stochastic flow* $(\Phi_t, 0 \le t \le T)$ *of smooth diffeomorphisms* $\mathbb{R}^n \to \mathbb{R}^n$ *associated to the equations* *(3.10).*

Proof. We refer to the book [Kunita (1990)], where the questions of existence and uniqueness of a smooth flow for stochastic differential equations driven by general continuous semimartingales are treated (cf. Theorem 3.4.1. p. 101 and Theorem 4.6.5. p. 173). $\square$

We consider now the following family of operators $(\mathcal{H}_T^N)_{T \ge 0}$ defined on the space of compactly supported smooth functions $f : \mathbb{R}^n \to \mathbb{R}$ by

$$(\mathcal{H}_T^N f)(x) = \mathbb{E}\left(f(X_T^x)\right), \ x \in \mathbb{R}^n.$$

The operator $\mathcal{H}_T^N$ shall be called the depth N holonomy operator.

Remark 3.16 *Of course, the operator* $\mathcal{H}_T^N$ *does not satisfy a semi-group property.*

Theorem 3.11 *Let* $f : \mathbb{R}^n \to \mathbb{R}$ *be a smooth, compactly supported function. In* $\mathbf{L}^2$,

$$\lim_{T \to 0} \frac{\mathcal{H}_T^N f - f}{T^{N+1}} = \Delta_N f,$$

where Δ_N *is a second order differential operator. It can be written*

$$\Delta_N = \sum_{i=1}^m Q_i^2$$

where:

(1)

$$m = \frac{1}{N+1} \sum_{k|N+1} \mu(k) d^{\frac{N+1}{k}};$$

(2) $Q_i \in \mathfrak{L}^{N+1}$ *is a universal Lie polynomial in* $V_1, ..., V_d$ *which is homogeneous of degree* $N + 1$.

Proof. Before we start the proof, let us precise some notations we already used. If $I = (i_1, ..., i_k) \in \{1, ..., d\}^k$ is a word, we denote $\mid I \mid = k$ its length and by V_I the commutator defined by

$$V_I = [V_{i_1}, [V_{i_2}, ..., [V_{i_{k-1}}, V_{i_k}]...].$$

The group of permutations of the index set $\{1, ..., k\}$ is denoted $\mathfrak{S}_k$. If $\sigma \in \mathfrak{S}_k$, we denote $e(\sigma)$ the cardinality of the set

$$\{j \in \{1, ..., k-1\}, \sigma(j) > \sigma(j+1)\}.$$

Finally, we denote

$$\Lambda_I(P_{\cdot,T}^N)_t = \sum_{\sigma \in \mathfrak{S}_k} \frac{(-1)^{e(\sigma)}}{k^2 \binom{k-1}{e(\sigma)}} \int_{0 \leq t_1 \leq ... \leq t_k \leq t} \circ dP_{t_1,T}^{N,\sigma^{-1}i_1} \circ ... \circ dP_{t_k,T}^{N,\sigma^{-1}i_k}.$$

Due to the scaling property

$$(P_{t,T}^N)_{0 \leq t \leq T} = (\sqrt{T} P_{\frac{t}{T},1}^N)_{0 \leq t \leq T},$$

we can closely follow the proof of Theorem 2.6, Chapter 2, to obtain the following asymptotic development of $f(X_T^x)$:

$$f(X_T^x) = \left(\exp \left(\sum_{k=1}^{2N+2} \sum_{I=(i_1,...,i_k)} \Lambda_I(P_{\cdot,T}^N)_T V_I \right) f \right) (x)$$

$$+ T^{\frac{2N+3}{2}} \mathbf{R}_{2N+3}(T, f, x),$$

where the remainder is bounded in probability when $T \to 0$. By definition of $(P_{t,T}^N)_{0 \leq t \leq T}$, we actually have

$$\sum_{k=1}^{2N+2} \sum_{I=(i_1,...,i_k)} \Lambda_I(P_{\cdot,T}^N)_T V_I = \sum_{k=N+1}^{2N+2} \sum_{I=(i_1,...,i_k)} \Lambda_I(P_{\cdot,T}^N)_T V_I,$$

so that

$$f(X_T^x) = \left(\exp \left(\sum_{k=N+1}^{2N+2} \sum_{I=(i_1,\ldots,i_k)} \Lambda_I(P_{.,T}^N)_T V_I \right) f \right)(x)$$
$$+ T^{\frac{2N+3}{2}} \mathbf{R}_{2N+3}(T,f,x).$$

Therefore,

$$\mathcal{H}_T^N f(x) = \mathbb{E} \left(\left(\exp \left(\sum_{k=N+1}^{2N+2} \sum_{I=(i_1,\ldots,i_k)} \Lambda_I(P_{.,T}^N)_T V_I \right) f \right)(x) \right)$$
$$+ T^{\frac{2N+3}{2}} \tilde{\mathbf{R}}_{2N+3}(T,f,x),$$

where

$$\tilde{\mathbf{R}}_{2N+1}(T,f,x) = \mathbb{E}\left(\mathbf{R}_{2N+1}(T,f,x)\right).$$

Since, by symmetry, we always have

$$\mathbb{E}\left(\Lambda_I(P_{.,T}^N)_T\right) = 0,$$

we have to go at the order 2 in the asymptotic development of the exponential when $T \to 0$. By neglecting the terms which have order more than $T^{\frac{2N+3}{2}}$, we obtain

$$\mathcal{H}_T^N f(x) = f(x) + \sum_{I,J} \frac{1}{2} \mathbb{E}\left(\Lambda_I(P_{.,T}^N)_T \Lambda_J(P_{.,T}^N)_T\right)(V_I V_J f)(x)$$
$$+ T^{\frac{2N+3}{2}} \mathbf{R}_{2N+3}^*(T,f,x),$$

where the remainder term $\mathbf{R}_{2N+3}^*(T,f,x)$ is bounded in $\mathbf{L}^2$ when $T \to 0$. Notice now that, since $(P_{t,T}^{i,N})_{0 \le t \le T} =^{law} (-P_{t,T}^{i,N})_{0 \le t \le T}$, we have

$$\mathbb{E}\left(\Lambda_I(P_{.,T}^N)_T \Lambda_J(P_{.,T}^N)_T\right) = 0,$$

as soon as $I \ne J$. Therefore

$$\mathcal{H}_T^N f(x) = f(x) + \sum_{I=(i_1,\ldots,i_{N+1})} \frac{1}{2} \mathbb{E}\left(\Lambda_I(P_{.,T}^N)_T^2\right)(V_I^2 f)(x)$$
$$+ T^{\frac{2N+3}{2}} \mathbf{R}_{2N+3}^*(T,f,x),$$

which leads to the expected result. $\qquad\qquad\square$

Example 3.15   We have

$$\Delta_0 = \frac{1}{2} \sum_{i=1}^{d} V_i^2,$$

and,

$$\Delta_1 = C \sum_{1 \le i < j \le d} [V_i, V_j]^2,$$

where C is a non negative constant.

Remark 3.17   *If $\mathcal{L}^{N+1} = 0$ then $\Delta_{N+1} = 0$ and if at some $x \in \mathbb{R}^n$, $\mathcal{L}^{N+1}(x) = \mathbb{R}^n$, then Δ_{N+1} satisfies the Hörmander's condition at x.*

Remark 3.18   *The convergence*

$$\frac{\mathcal{H}_T^{N+1} - Id}{T^{N+1}} \to_{T \to 0} \Delta_{N+1},$$

also holds in $\mathbf{L}^p$, $p \ge 1$.

Theorem 3.12   *Assume that $\mathcal{L}^{N+1} = 0$, then for any solution $(X_t^{x_0})_{0 \le t \le T}$ of (3.10) we have almost surely $X_T^{x_0} = x_0$. On the other hand, assume that $\mathcal{L}^{N+1}(x_0) = \mathbb{R}^n$, then for the solution $(X_t^{x_0})_{0 \le t \le T}$ of (3.10) the random variable $X_T^{x_0}$ has a smooth density with respect to the Lebesgue measure of $\mathbb{R}^n$.*

Proof.   Assume that $\mathcal{L}^{N+1} = 0$, then a similar argument as those given in the proof of Theorem 2.5 in Chapter 2, there exists a smooth map

$$F : \mathbb{R}^n \times \mathbf{G}_N(\mathbb{R}^d) \to \mathbb{R}^n$$

such that, for $x_0 \in \mathbb{R}^n$, the solution $(X_t^{x_0})_{0 \le t \le T}$ of the SDE (3.10) can be written

$$X_t^{x_0} = F(x_0, Q_{t,T}^N),$$

which implies immediately the expected result.

Assume now that $\mathcal{L}^{N+1}(x_0) = \mathbb{R}^n$. Let us consider the solution $(Z_t)_{t \ge 0}$ of the following stochastic differential equation:

$$Z_t = x_0 + \sum_{i=1}^{d} \int_0^t V_i(Z_s) \circ dB_s^i, \ t \ge 0,$$

where $(B_t^1, ..., B_t^d)_{t \geq 0}$ is a d-dimensional standard Brownian motion. From theorem 3.3, the random variable

$$(Z_T, B_T^*)$$

has a smooth density with respect to any Lebesgue measure on $\mathbb{R}^n \times \mathbb{G}_N(\mathbb{R}^d)$. It implies the existence of a smooth function $p : \mathbb{R}^n \to \mathbb{R}$ such that for every bounded measurable function $f : \mathbb{R}^n \to \mathbb{R}$

$$\mathbb{E}(f(Z_T) \mid B_T^* = 0) = \int_{\mathbb{R}^n} f(y)p(y)dy.$$

Now, since in law the process $(P_{t,T}^N)_{0 \leq t \leq T}$, is identical to the Brownian motion $(B_t)_{0 \leq t \leq T}$ conditioned by $B_T^* = 0$, the function p is actually exactly the density of the random variable $X_T^{x_0}$ where $(X_t^{x_0})_{0 \leq t \leq T}$ is the solution of (3.10) with initial condition x_0. $\quad\square$

Remark 3.19 *In the case $N = 1$, we can give another proof of the existence of a density under the assumption $\mathcal{L}^2(x_0) = \mathbb{R}^n$. Indeed, let us recall that*

$$P_{t,T}^1 = (T - t) \int_0^t \frac{dB_s}{T - s}, \quad t < T, \text{ and } P_{T,T} = 0.$$

Using this formula, it is not difficult to prove that if $(X_t)_{0 \leq t \leq T}$ is a solution of (3.10), then $X_T \in \mathbb{D}^\infty$ (see Appendix A, the section about Malliavin calculus for the definition of this space). Moreover, a direct computation shows that for any $0 \leq s \leq T$,

$$\mathbf{D}_s X_T = \mathbf{J}_{0 \to T} \left(\mathbf{J}_{0 \to s}^{-1} \sigma(X_s) - \frac{1}{T - s} \int_s^T \mathbf{J}_{0 \to u}^{-1} \sigma(X_u)du \right),$$

where $(\mathbf{J}_{0 \to t})_{0 \leq t \leq T}$ is the first variation process defined by

$$\mathbf{J}_{0 \to t} = \frac{\partial \Phi_t}{\partial x},$$

and σ the $n \times d$ matrix field $\sigma = (V_1, ..., V_d)$. From this, we can deduce that the Malliavin matrix of X_T must be invertible. Indeed, if not, we could find a non zero vector $h \in \mathbb{R}^d$ such that $\mathbf{D}_s X_T \cdot h = 0$ for $0 \leq s \leq T$, this leads to the conclusion that $(\mathbf{J}_{0 \to s}^{-1} \sigma(X_s) \cdot h)_{0 \leq s \leq T}$ must be constant. An iterative application of Itô's formula, as performed in the proof of Hörmander's theorem given at the beginning of the chapter, shows then that h must be orthogonal to $\mathcal{L}^2(x_0)$ which leads to the expected contradiction. Thus, X_T admits a density with respect to the Lebesgue measure.

Remark 3.20   *In general, the only condition $\mathcal{L}^{N+1}(x_0) = 0$ is not enough to conclude that for the solution $(X_t^{x_0})_{0 \leq t \leq T}$ of (3.10) we have almost surely $X_T^{x_0} = x_0$. For instance for $N = 2$, consider in dimension 2,*

$$V_1 = \begin{pmatrix} 1 \\ 0 \end{pmatrix}, \text{ and } V_2 = \begin{pmatrix} 0 \\ f(x) \end{pmatrix},$$

where f is a smooth function whose Taylor development at 0 is 0 (e.g. $f(x) = e^{-\frac{1}{x^2}} 1_{x>0}$). Nevertheless, if the vector fields V_i's are assumed to be analytic, it is true that $\mathcal{L}^{N+1}(x_0) = 0$ implies that almost surely $X_T^{x_0} = x_0$.

In the case of the existence of a density for $X_T^{x_0}$, we can moreover give an equivalent of this density when the length of the loop tends to 0. To this end, let us precise some notations.

We set for $x \in \mathbb{R}^n$ and $k \geq N$,

$$\mathcal{U}_k(x) = \text{span}\{V_I, \ N \leq |I| \leq k\}.$$

In the case where $\mathcal{L}^{N+1}(x) = \mathbb{R}^n$, if k is big enough then $\mathcal{U}_k(x) = \mathbb{R}^n$. We denote $d(x)$ the smallest integer $k \geq N+1$ for which this equality holds and define the graded dimension

$$\dim_{\mathcal{H}} \mathcal{L}^{N+1}(x) := \sum_{k=N+1}^{d(x)} k \left(\dim \mathcal{U}_k(x) - \dim \mathcal{U}_{k-1}(x) \right).$$

Theorem 3.13   *Assume that for any $x \in \mathbb{R}^n$, $\mathcal{L}^{N+1}(x) = \mathbb{R}^n$. Let us denote $p_T(x)$ the density of X_T^x with respect to the Lebesgue measure. We have*

$$p_T(x) \sim_{T \to 0} \frac{m(x)}{T^{\frac{\dim_{\mathcal{H}} \mathcal{L}^{N+1}(x)}{2}}},$$

where m is a smooth non negative function.

Proof.   Let us, once time again, consider the solution $(Z_t^x)_{t \geq 0}$ of the following stochastic differential equation:

$$Z_t^x = x + \sum_{i=1}^{d} \int_0^t V_i(Z_s^x) \circ dB_s^i, \ t \geq 0,$$

where $(B_t^1, ..., B_t^d)_{t \geq 0}$ is a d-dimensional standard Brownian motion. From Corollary 3.2 , the density at $(x, 0)$ of the random variable (Z_T^x, B_T^*) behaves when T goes to zero like

$$\frac{\tilde{m}(x)}{T^{\frac{\dim_{\mathcal{H}} \mathbb{G}_N(\mathbb{R}^d) + \dim_{\mathcal{H}} \mathcal{L}^{N+1}(x)}{2}}},$$

where $\dim_{\mathcal{H}} \mathbb{G}_N(\mathbb{R}^d) = \sum_{j=1}^N j \dim \mathcal{V}_j$ is the graded dimension of $\mathbb{G}_N(\mathbb{R}^d)$, and $\tilde{m}$ a smooth non negative function. Always from Corollary 3.2, the density of the random variable Y_T behaves when T goes to zero like

$$\frac{C}{T^{\frac{\dim_{\mathcal{H}} \mathbb{G}_N(\mathbb{R}^d)}{2}}},$$

where C is a non negative constant. The conclusion follows readily. $\square$

Appendix A

Basic Stochastic Calculus

In this Appendix, we review briefly the basic theory of stochastic processes and stochastic differential equations. The stochastic integration is a natural, easy and fruitful integration theory which is due to [Itô (1944)]. For a much more complete exposition of the stochastic calculus we refer to [Dellacherie and Meyer (1976)], [McKean (1969)], [Protter (2004)], and [Revuz and Yor (1999)]. For further reading on stochastic differential equations we refer to [Elworthy (1982)], [Ikeda and Watanabe (1989)], [Kunita (1990)], [Stroock (1982)], and [Stroock and Varadhan (1979)]. We assume some familiarity with basic probability theory.

A.1 Stochastic processes and Brownian motion

Let $(\Omega, (\mathcal{F}_t)_{t \geq 0}, \mathcal{F}, \mathbb{P})$ be a filtered probability space which satisfies the usual conditions, that is:

(1) $(\mathcal{F}_t)_{t \geq 0}$ is a filtration, i.e. an increasing family of sub-σ-fields of $\mathcal{F}$;
(2) for any $t \geq 0$, $\mathcal{F}_t$ is complete with respect to $\mathbb{P}$, i.e. every subset of a set of measure zero is contained in $\mathcal{F}_t$;
(3) $(\mathcal{F}_t)_{t \geq 0}$ is right continuous, i.e. for any $t \geq 0$,

$$\mathcal{F}_t = \bigcap_{s > t} \mathcal{F}_s.$$

A stochastic process $(X_t)_{t \geq 0}$ on $(\Omega, (\mathcal{F}_t)_{t \geq 0}, \mathcal{F}, \mathbb{P})$ is an application

$$\mathbb{R}_{\geq 0} \times \Omega \to \mathbb{R}$$
$$(t, \omega) \quad \to X_t(\omega),$$

which is measurable with respect to $\mathcal{B}(\mathbb{R}_{\geq 0}) \otimes \mathcal{F}$. The process $(X_t)_{t \geq 0}$ is said to be adapted (with respect to the filtration $(\mathcal{F}_t)_{t \geq 0}$) if for every $t \geq 0$, X_t is $\mathcal{F}_t$-measurable. It is said to be continuous if for almost every $\omega \in \Omega$, the function $t \to X_t(\omega)$ is continuous. A stochastic process $(\tilde{X}_t)_{t \geq 0}$ is called a modification of $(X_t)_{t \geq 0}$ if for every $t \geq 0$,

$$\mathbb{P}\left(X_t = \tilde{X}_t\right) = 1.$$

The following theorem known as Kolmogorov's continuity criterion is fundamental.

Theorem A.1 *Let* $\alpha, \varepsilon, c > 0$. *If a process* $(X_t)_{t \geq 0}$ *satisfies for every* $s, t \geq 0$,

$$\mathbb{E}\left(\mid X_t - X_s \mid^\alpha\right) \leq c \mid t - s \mid^{1 + \varepsilon},$$

then there exists a modification of $(X_t)_{t \geq 0}$ *which is a continuous process.*

One of the most important examples of stochastic process is the Brownian motion. A process $(B_t)_{t \geq 0}$ defined on $(\Omega, (\mathcal{F}_t)_{t \geq 0}, \mathbb{P})$ is said to be a standard Brownian motion with respect to the filtration $(\mathcal{F}_t)_{t \geq 0}$ if:

(1) $B_0 = 0$ a.s.;
(2) $B_1 =^{law} \mathcal{N}(0,1)$;
(3) $(B_t)_{t \geq 0}$ is $\mathcal{F}$-adapted;
(4) For any $t > s \geq 0$,

$$B_t - B_s =^{law} B_{t-s};$$

(5) For any $t > s \geq 0$, $B_t - B_s$ is independent of $\mathcal{F}_s$.

If a process is a Brownian motion with respect to its own filtration, we simply say that it is a standard Brownian motion, without mentioning the underlying filtration. A d-dimensional process $(B_t)_{t \geq 0} = (B_t^1, ..., B_t^d)_{t \geq 0}$ is said to be a d-dimensional standard Brownian motion if the processes $(B_t^1)_{t \geq 0}, ..., (B_t^d)_{t \geq 0}$ are independent standard Brownian motions.

A standard Brownian motion $(B_t)_{t \geq 0}$ is a Gaussian process, that is for every $n \in \mathbb{N}^*$, and $0 \leq t_1 \leq ... \leq t_n$, the random variable

$$(B_{t_1}, ..., B_{t_n})$$

is Gaussian. The mean function of $(B_t)_{t \geq 0}$ is

$$\mathbb{E}(B_t) = 0,$$

and its covariance function is

$$\mathbb{E}(B_s B_t) = \inf(s, t).$$

It is easy to show, thanks to Kolmogorov's continuity criterion, that it is always possible to find a continuous modification of a standard Brownian motion. Of course, we always work with this continuous modification. The law of a standard Brownian motion, which therefore lives on the space of continuous functions $\mathbb{R}_{\geq 0} \to \mathbb{R}$ is called the Wiener measure. If $(B_t)_{t \geq 0}$ is a standard Brownian motion, then the following properties hold:

(1) For every $c > 0$, $(B_{ct})_{t \geq 0} =^{law} (\sqrt{c} B_t)_{t \geq 0}$;
(2) $(t B_{\frac{1}{t}})_{t \geq 0} =^{law} (B_t)_{t \geq 0}$;
(3) For almost every $\omega \in \Omega$, the path $t \to B_t(\omega)$ is nowhere differentiable and locally Hölder continuous of order α for every $\alpha < \frac{1}{2}$;
(4) For every subdivision $0 = t_0 \leq \cdots \leq t_n = t$ whose mesh tends to 0, almost surely we have

$$\lim_{n \to +\infty} \sum_{i=0}^{n-1} \left(B_{t_{i+1}} - B_{t_i} \right)^2 = t.$$

A.2   Markov processes

Intuitively, a Markov process $(X_t)_{t \geq 0}$ defined on a filtered probability space $(\Omega, (\mathcal{F}_t)_{t \geq 0}, \mathcal{F}, \mathbb{P})$ is a process without memory.

Definition A.1   A transition function $\{P_t, t \geq 0\}$ on $\mathbb{R}$ is a family of kernels $P_t : \mathbb{R} \times \mathcal{B}(\mathbb{R}) \to [0, 1]$ such that:

(1) For $t \geq 0$ and $x \in \mathbb{R}$, $P_t(x, \cdot)$ is a Borel measure on $\mathcal{B}(\mathbb{R})$ with $P_t(x, \mathbb{R}) \leq 1$;
(2) For $t \geq 0$ and $A \in \mathcal{B}(\mathbb{R})$, the function $x \to P_t(x, A)$ is measurable with respect to $\mathcal{B}(\mathbb{R})$;
(3) For $s, t \geq 0$, $x \in \mathbb{R}$ and $A \in \mathcal{B}(\mathbb{R})$,

$$P_{t+s}(x, A) = \int_{\mathbb{R}} P_t(y, A) P_s(x, dy); \tag{A.1}$$

Equation (A.1) is called the Chapman-Kolmogorov equation. We may equally think of the transition function as being a family $(P_t)_{t \geq 0}$ of positive bounded operators of norm less or equal to 1 on the space of bounded Borel functions $f : \mathbb{R} \to \mathbb{R}$ defined by

$$(P_t f)(x) = \int_{\mathbb{R}} f(y) P_t(x, dy).$$

Observe that the Chapman-Kolmogorov equation is equivalent to the semi-group property

$$P_{t+s} = P_t P_s.$$

Definition A.2 An adapted continuous process $(X_t)_{t \geq 0}$ defined on $(\Omega, (\mathcal{F}_t)_{t \geq 0}, \mathcal{F}, \mathbb{P})$ is said to be a Markov process (with respect to the filtration $(\mathcal{F}_t)_{t \geq 0}$) if there exists a transition function $(P_t)_{t \geq 0}$ such that for every bounded Borel functions $f : \mathbb{R} \to \mathbb{R}$, $0 \leq s \leq t$,

$$\mathbb{E}\left(f(X_{s+t}) \mid \mathcal{F}_s\right) = (P_t f)(X_s).$$

In the theory of stochastic processes, it is often useful to deal with random times. A real valued positive random variable T is said to be a stopping time with respect to the filtration $(\mathcal{F}_t)_{t \geq 0}$ if for any $t \geq 0$,

$$\{T \leq t\} \in \mathcal{F}_t.$$

If T is a stopping time the smallest σ-field which contains all the events $\{T \leq t\}$, $t \geq 0$, is denoted $\mathcal{F}_T$.

Definition A.3 An adapted continuous process $(X_t)_{t \geq 0}$ defined on $(\Omega, (\mathcal{F}_t)_{t \geq 0}, \mathcal{F}, \mathbb{P})$ is said to be a strong Markov process (with respect to the filtration $(\mathcal{F}_t)_{t \geq 0}$) if there exists a transition function $(P_t)_{t \geq 0}$ such that for every bounded Borel functions $f : \mathbb{R} \to \mathbb{R}$, $t \geq 0$,

$$\mathbb{E}\left(f(X_{S+t}) \mid \mathcal{F}_S\right) = (P_t f)(X_S),$$

where S is any stopping time that satisfies almost surely $0 \leq S \leq t$.

Observe that a strong Markov process is always a Markov process and that a standard Brownian motion is a strong Markov process. For further details on Markov processes, we refer to Chapter III of [Rogers and Williams (2000)].

A.3 Martingales

Consider an adapted and continuous process $(M_t)_{t \geq 0}$ defined on a filtered probability space $(\Omega, (\mathcal{F}_t)_{t \geq 0}, \mathcal{F}, \mathbb{P})$.

Definition A.4 The process $(M_t)_{t\geq 0}$ is said to be a martingale (with respect to the filtration $(\mathcal{F}_t)_{t\geq 0}$) if:

(1) For $t \geq 0$, $\mathbb{E}(|M_t|) < +\infty$;
(2) For $0 \leq s \leq t$, $\mathbb{E}(M_t \mid \mathcal{F}_s) = M_s$.

For instance a standard Brownian motion is a martingale. For martingales, we have the following proposition, which is known as the stopping theorem.

Proposition A.1 *The following properties are equivalent:*

(1) The process $(M_t)_{t\geq 0}$ is a martingale;
(2) For any bounded stopping time T, $\mathbb{E}(M_T) = \mathbb{E}(M_0)$;
(3) For any pair of bounded stopping times S and T, with $S \leq T$,
   $\mathbb{E}(M_T \mid \mathcal{F}_S) = M_S.$

Actually, if T is a bounded stopping time, then the process $(M_{t\wedge T})_{t\geq 0}$ is also a martingale.

A martingale $(M_t)_{t\geq 0}$ is said to be square integrable if for $t \geq 0$, $\mathbb{E}(M_t^2) < +\infty$. In that case, from Jensen's inequality the function $t \to \mathbb{E}(M_t^2)$ is increasing. We also have the so-called Doob's inequality

$$\mathbb{E}\left(\sup_{t\geq 0} M_t^2\right) \leq 4 \sup_{t\geq 0} \mathbb{E}(M_t^2),$$

which is one of the cornerstone of the stochastic integration. Observe therefore that if $\sup_{t\geq 0} \mathbb{E}(M_t^2) < +\infty$, the martingale $(M_t)_{t\geq 0}$ is uniformly integrable and converges in L^2 to a square integrable random variable M_∞ which satisfies

$$\mathbb{E}(M_\infty \mid \mathcal{F}_t) = M_t, \ t \geq 0.$$

If $(M_t)_{t\geq 0}$ is a square integrable martingale, there exists a unique increasing process denoted $(\langle M\rangle_t)_{t\geq 0}$ which satisfies:

(1) $\langle M\rangle_0 = 0$;
(2) The process $(M_t^2 - \langle M\rangle_t)_{t\geq 0}$ is a martingale.

This increasing process $(\langle M\rangle_t)_{t\geq 0}$ is called the quadratic variation of the martingale $(M_t)_{t\geq 0}$. This terminology comes from the following property. If $0 = t_0 \leq t_1 \leq \cdots \leq t_n = t$ is a subdivision of the time interval $[0, t]$

whose mesh tends to 0, then in probability

$$\lim_{n \to +\infty} \sum_{i=0}^{n-1} \left(M_{t_{i+1}} - M_{t_i} \right)^2 = \langle M \rangle_t.$$

For technical reasons (localization procedures), we often have to consider a wider class than martingales.

Definition A.5   The process $(M_t)_{t \geq 0}$ is said to be a local martingale (with respect to the filtration $(\mathcal{F}_t)_{t \geq 0}$) if there exists a sequence $(T_n)_{n \geq 0}$ of stopping times such that:

(1) The sequence $(T_n)_{n \geq 0}$ is increasing and $\lim_{n \to +\infty} T_n = +\infty$ almost surely;

(2) For every $n \geq 1$, the process $(M_{t \wedge T_n})_{t \geq 0}$ is a uniformly integrable martingale with respect to the filtration $(\mathcal{F}_t)_{t \geq 0}$.

A martingale is always a local martingale but the converse is not true. Nevertheless, a local martingale $(M_t)_{t \geq 0}$ such that for every $t \geq 0$,

$$\mathbb{E} \left(\sup_{s \leq t} |M_s| \right) < +\infty,$$

is a martingale. If $(M_t)_{t \geq 0}$ is a local martingale, there still exists a unique increasing process denoted $(\langle M \rangle_t)_{t \geq 0}$ which satisfies:

(1) $\langle M \rangle_0 = 0$;

(2) The process $(M_t^2 - \langle M \rangle_t)_{t \geq 0}$ is a local martingale.

This increasing process $(\langle M \rangle_t)_{t \geq 0}$ is called the quadratic variation of the local martingale $(M_t)_{t \geq 0}$. By polarization, it is easily seen that, more generally, if $(M_t)_{t \geq 0}$ and $(N_t)_{t \geq 0}$ are two continuous local martingales, then there exists a unique continuous process denoted $(\langle M, N \rangle_t)_{t \geq 0}$ and called the quadratic covariation of $(M_t)_{t \geq 0}$ and $(N_t)_{t \geq 0}$ which satisfies:

(1) $\langle M, N \rangle_0 = 0$;

(2) The process $(M_t N_t - \langle M, N \rangle_t)_{t \geq 0}$ is a local martingale.

Before we turn to the theory of stochastic integration, we conclude this section with Lévy's characterization of Brownian motion.

Proposition A.2   *Let $(M_t)_{t \geq 0}$ be a d-dimensional continuous local martingale such that $M_0 = 0$ and*

$$\langle M^i, M^i \rangle_t = t, \quad \langle M^i, M^j \rangle_t = 0 \ \text{if } i \neq j.$$

Then $(M_t)_{t \geq 0}$ is a standard Brownian motion.

A.4 Stochastic integration

Let $(\Omega, (\mathcal{F}_t)_{t \geq 0}, \mathcal{F}, \mathbb{P})$ be a filtered probability space which satisfies the usual conditions specified before. We aim now at defining an integral $\int_0^t H_s dM_s$ where $(M_t)_{t \geq 0}$ is an adapted continuous square integrable martingale such that $\sup_{t \geq 0} \mathbb{E}\left(M_t^2\right) < +\infty$ and $(H_t)_{t \geq 0}$ an adapted process which shall be in a *good* class. Observe that such an integral could not be defined trivially since the Young's integration theory does not cover the integration against paths which are less than $\frac{1}{2}$-Hölder continuous. First, we define the class of integrands. The predictable σ-field $\mathcal{P}$ associated with the filtration $(\mathcal{F}_t)_{t \geq 0}$ is the σ-field generated on $\mathbb{R}_{\geq 0} \times \Omega$ by the space of indicator functions $\mathbf{1}_{]S,T]}$, where S and T are two stopping times such that $S \leq T$. An adapted stochastic process $(H_t)_{t \geq 0}$ is said to be predictable if the application $(t, \omega) \rightarrow H_t(\omega)$ is measurable with respect to $\mathcal{P}$. Observe that a continuous adapted process is predictable.

Let us first assume that $(H_t)_{t \geq 0}$ is a predictable elementary process that can be written

$$H_t = \sum_{i=1}^{n-1} H_i \mathbf{1}_{]T_i, T_{i+1}]}(t)$$

where H_i is a $\mathcal{F}_{T_i}$ measurable bounded random variable, and where $(T_i)_{1 \leq i \leq n}$ is a finite increasing sequence of stopping times. In that case, a natural definition for $\int_0^t H_s dM_s$ is

$$\int_0^t H_s dM_s = \sum_{i=1}^{n-1} H_i \left(M_{t \wedge T_{i+1}} - M_{t \wedge T_i}\right).$$

Then, we observe that the process $\left(\int_0^t H_s dM_s\right)_{t \geq 0}$ is a bounded martingale which satisfies furthermore from Doob's inequality

$$\mathbb{E}\left(\sup_{t \geq 0} \left(\int_0^t H_s dM_s\right)^2\right) \leq 4 \parallel H \parallel_\infty^2 \mathbb{E}(M_\infty^2).$$

We also note that

$$\mathbb{E}\left(\left(\int_0^t H_s dM_s\right)^2\right) = \mathbb{E}\left(\sum_{i=1}^{n-1} H_i \left(M_{t \wedge T_{i+1}} - M_{t \wedge T_i}\right)^2\right).$$

Assume now that $(H_t)_{t \geq 0}$ is a bounded continuous adapted process. To define $\int_0^t H_s dM_s$, the idea is of course to approximate $(H_t)_{t \geq 0}$ with elementary processes $(H_t^p)_{t \geq 0}$ and to check the convergence of $\left(\int_0^t H_s^p dM_s \right)_{t \geq 0}$ with respect to a suitable norm. Precisely, let us define for any $p \in \mathbb{N}^*$, the following stopping times:

$$T_0^p = 0$$

$$T_1^p = \inf \left\{ t > 0, |\, H_t \,| \geq \frac{1}{p} \right\},$$

and by iteration

$$T_{n+1}^p = \inf \left\{ t > T_n^p, |\, H_t - H_{T_n^p} \,| \geq \frac{1}{p} \right\}.$$

We now define

$$H_t^p = \sum_{i=1}^{n-1} H_{T_i^p} \mathbf{1}_{]T_i^p, T_{i+1}^p]}(t).$$

For this sequence of processes $(H_t^p)_{t \geq 0}$, it easy to show that

$$\mathbb{E} \left(\sup_{t \geq 0} \left(\int_0^t (H_s^p - H_s^q) dM_s \right)^2 \right) \to_{p,q \to +\infty} 0.$$

From this, we can deduce that there exists a continuous martingale denoted $\left(\int_0^t H_s dM_s \right)_{t \geq 0}$ such that

$$\int_0^t H_s dM_s = \lim_{p \to +\infty} \int_0^t H_s^p dM_s$$

uniformly for t on compact sets. We furthermore have

$$\mathbb{E} \left(\sup_{t \geq 0} \left(\int_0^t H_s dM_s \right)^2 \right) \leq 4 \, \| H \|_\infty^2 \, \mathbb{E}(M_\infty^2).$$

and

$$\mathbb{E} \left(\left(\int_0^t H_s dM_s \right)^2 \right) = \mathbb{E} \left(\int_0^t H_s^2 d\langle M \rangle_s \right).$$

Now, by localization, it is not difficult to extend naturally the definition of $\int_0^t H_s dM_s$ in the general case where:

(1) $(M_t)_{t \geq 0}$ is a semimartingale, that is, $(M_t)_{t \geq 0}$ can be written

$$M_t = A_t + N_t,$$

where $(A_t)_{t \geq 0}$ is a bounded variation process and $(N_t)_{t \geq 0}$ is a local martingale;

(2) $(H_t)_{t \geq 0}$ is a locally bounded predictable process.

In this setting, we have:

$$\int_0^t H_s dM_s = \int_0^t H_s dA_s + \int_0^t H_s dN_s.$$

Observe that, since $(A_t)_{t \geq 0}$ is a bounded variation process, the integral $\int_0^t H_s dA_s$ is simply a Riemann-Stieltjes integral. The process $\left(\int_0^t H_s dN_s \right)_{t \geq 0}$ is a local martingale.

The class of semimartingales appears then as a good class of integrators in the theory of stochastic integration. It can be shown that this is actually the widest possible class if we wish to obtain a *natural* integration theory (see Dellacherie-Meyer [Dellacherie and Meyer (1976)] or Protter [Protter (2004)] for a precise statement). The decomposition of a semimartingale $(M_t)_{t \geq 0}$ under the form

$$M_t = A_t + N_t,$$

is essentially unique under the condition $N_0 = 0$. The process $(A_t)_{t \geq 0}$ is called the bounded variation part of $(M_t)_{t \geq 0}$. The process $(N_t)_{t \geq 0}$ is called the local martingale part of $(M_t)_{t \geq 0}$. If $(M_t^1)_{t \geq 0}$ and $(M_t^2)_{t \geq 0}$ are two semimartingales, then we define the quadratic covariation $\left(\langle M^1, M^2 \rangle_t \right)_{t \geq 0}$ of $(M_t^1)_{t \geq 0}$ and $(M_t^2)_{t \geq 0}$ as being $\left(\langle N^1, N^2 \rangle_t \right)_{t \geq 0}$ where $(N_t^1)_{t \geq 0}$ and $(N_t^2)_{t \geq 0}$ are the local martingale parts.

Throughout this book we shall only deal with continuous processes so that in the sequel, we shall often omit to precise the continuity of the processes which will be considered. Moreover, we shall preferably use Stratonovitch's integration rather than Itô's integration. If $(N_t)_{0 \leq t \leq T}$, $T > 0$, is an $\mathcal{F}$-adapted real valued local martingale and if $(\Theta_t)_{0 \leq t \leq T}$ is an $\mathcal{F}$-adapted continuous process satisfying $\mathbb{E} \left(\int_0^T \Theta_t^2 d\langle N \rangle_t \right)$, then by definition,

$$\int_0^T \Theta_t \circ dN_t = \int_0^T \Theta_t \cdot dN_t + \frac{1}{2} \langle \Theta, N \rangle_T,$$

where:

(1) $\int_0^T \Theta_t \circ dN_t$ is the Stratonovitch integral of $(\Theta_t)_{0 \leq t \leq T}$ against $(N_t)_{0 \leq t \leq T}$;

(2) $\int_0^T \Theta_t \cdot dN_t$ is the Itô integral of $(\Theta_t)_{0 \leq t \leq T}$ against $(N_t)_{0 \leq t \leq T}$;

(3) $\langle \Theta, N \rangle_T$ is the quadratic covariation at time T between $(\Theta_t)_{0 \leq t \leq T}$ and $(N_t)_{0 \leq t \leq T}$.

A.5   Itô's formula

The Itô's formula is certainly the most important formula of stochastic calculus.

Theorem A.2   *Let* $(X_t)_{t \geq 0} = \left(X_t^1, \cdots, X_t^n \right)_{t \geq 0}$ *be a n- dimensional continuous semimartingale. Let now* $f : \mathbb{R}^n \to \mathbb{R}$ *be a* C^2 *function. We have*

$$f(X_t) = f(X_0) + \sum_{i=1}^n \int_0^t \frac{\partial f}{\partial x_i}(X_s) dX_s^i + \frac{1}{2} \sum_{i,j=1}^n \int_0^t \frac{\partial^2 f}{\partial x_i \partial x_j}(X_s) d\langle X^i, X^j \rangle_s$$

$$= f(X_0) + \sum_{i=1}^n \int_0^t \frac{\partial f}{\partial x_i}(X_s) \circ dX_s^i.$$

A.6   Girsanov's theorem

We give here the most important theorem concerning the impact of a change of probability measure on the class of semimartingales. Let $\mathcal{C}(\mathbb{R}_{\geq 0}, \mathbb{R}^d)$ denote the space of continuous functions $\mathbb{R}_{\geq 0} \to \mathbb{R}^d$. Let X_s, $s \geq 0$, denote the coordinate mappings, that is

$$X_s(\omega) = \omega(s), \quad \omega \in \mathcal{C}(\mathbb{R}_{\geq 0}, \mathbb{R}^d).$$

Set now $\mathcal{B}_t^0 = \sigma(X_s, s \leq t)$ and consider on $\mathcal{C}(\mathbb{R}_{\geq 0}, \mathbb{R}^d)$ the Wiener measure $\mathbb{P}$, that is the law of a d-dimensional standard Brownian motion. Let finally $\mathcal{B}_t$ be the usual $\mathbb{P}$-augmentation of $\mathcal{B}_t^0$ and

$$\mathcal{B}_\infty = \vee_{t \geq 0} \mathcal{B}_t.$$

The filtered probability space

$$\left(\mathcal{C}(\mathbb{R}_{\geq 0}, \mathbb{R}^d), (\mathcal{B}_t)_{t \geq 0}, \mathcal{B}_\infty, \mathbb{P} \right)$$

is called the Wiener space. Observe that, by definition of $\mathbb{P}$, the process $(X_t)_{t\geq 0}$ is under $\mathbb{P}$ a d-dimensional standard Brownian motion. We have the following theorem, referred to as the Girsanov's theorem.

Theorem A.3

(1) *Let $\mathbb{Q}$ be a probability measure on $\left(\mathcal{C}(\mathbb{R}_{\geq 0}, \mathbb{R}^d), \mathcal{B}_\infty\right)$ which is equivalent to $\mathbb{P}$. Let us denote by D the density of $\mathbb{Q}$ with respect to $\mathbb{P}$. Then there exists an adapted continuous $\mathbb{R}^d$-valued process $(\Theta_t)_{t\geq 0}$ such that*

$$\mathbb{E}\left(D \mid \mathcal{B}_t\right) = \exp\left(\int_0^t \Theta_s dX_s - \frac{1}{2}\int_0^t \|\Theta_s\|^2 ds\right),$$

and under $\mathbb{Q}$,

$$\tilde{X}_t = X_t - \int_0^t \Theta_s ds$$

is a standard Brownian motion.

(2) *Let $(\Theta_t)_{t\geq 0}$ be an adapted continuous $\mathbb{R}^d$-valued process such that the process*

$$Z_t = \exp\left(\int_0^t \Theta_s dX_s - \frac{1}{2}\int_0^t \|\Theta_s\|^2 ds\right), \quad t \geq 0,$$

is a uniformly integrable martingale under $\mathbb{P}$. Define a probability measure on $\left(\mathcal{C}(\mathbb{R}_{\geq 0}, \mathbb{R}^d), \mathcal{B}_\infty\right)$ by:

$$d\mathbb{Q} = Z_\infty \mathbb{P}.$$

Then, under $\mathbb{Q}$, the process

$$X_t - \int_0^t \Theta_s ds$$

is a standard Brownian motion.

Observe that the process

$$Z_t = \exp\left(\int_0^t \Theta_s dX_s - \frac{1}{2}\int_0^t \|\Theta_s\|^2 ds\right), \quad t \geq 0,$$

satisfies the following equation

$$Z_t = 1 + \int_0^t Z_s \Theta_s dX_s.$$

It can be shown that sufficient conditions which ensure that $(Z_t)_{t\geq 0}$ is a uniformly integrable martingale are the following:

(1) For any $t \geq 0$,

$$\mathbb{E}(Z_t) = 1;$$

(2) For any $t \geq 0$,

$$\mathbb{E}\left(\exp\left(\frac{1}{2}\int_0^t \|\Theta_s\|^2 \, ds\right)\right) < +\infty.$$

A.7   Stochastic differential equations

Let $(\Omega, (\mathcal{F}_t)_{t\geq 0}, \mathbb{P})$ be a filtered probability space which satisfies the usual conditions. Let

$$(M_t)_{t\geq 0} = (M_t^1, ..., M_t^d)_{t\geq 0}$$

denote a continuous d-dimensional semimartingale that is adapted to the filtration $(\mathcal{F}_t)_{t\geq 0}$. Consider now d smooth vector fields $V_i : \mathbb{R}^n \to \mathbb{R}^n$, $n \geq 1$, $i = 1, ..., d$. The fundamental theorem for the existence and the uniqueness of solutions for stochastic differential equations is the following:

Theorem A.4   *Let $x_0 \in \mathbb{R}^n$. On $(\Omega, (\mathcal{F}_t)_{t\geq 0}, \mathbb{P})$, there exists a unique continuous process $(X_t^{x_0})_{t\geq 0}$ adapted to the natural filtration of $(M_t)_{t\geq 0}$ completed with respect to $\mathbb{P}$ and such that for $t \geq 0$,*

$$X_t^{x_0} = x_0 + \sum_{i=1}^d \int_0^t V_i(X_s^{x_0}) \circ dM_s^i. \tag{A.2}$$

The process $(X_t^{x_0})_{t\geq 0}$ is called the solution of the stochastic differential equation (A.2).

In the sequel the stochastic differential equation (A.2) shall be denoted $\mathbf{SDE}\,(x_0, (V_i)_{1\leq i\leq d}, (M_t)_{t\geq 0})$. Observe that $\mathbf{SDE}\,(x_0, (V_i)_{1\leq i\leq d}, (M_t)_{t\geq 0})$ is written in Stratonovitch's form. Thanks to Itô's formula the corresponding Itô's formulation is

$$X_t^{x_0} = x_0 + \frac{1}{2}\sum_{i,j=1}^d \int_0^t \nabla_{V_i} V_j(X_s^{x_0}) d\langle M_s^i, M_s^j\rangle + \sum_{i=1}^d \int_0^t V_i(X_s^{x_0}) dM_s^i,$$

where for $1 \leq i, j \leq d$, $\nabla_{V_i} V_j$ is the vector field given by

$$\nabla_{V_i} V_j(x) = \sum_{k=1}^{n} \left(\sum_{l=1}^{n} v_l^k(x) \frac{\partial v_l^j}{\partial x_l}(x) \right) \frac{\partial}{\partial x_k}, \quad x \in \mathbb{R}^n.$$

An important point is that the solution $(X_t^{x_0})_{t\geq0}$ of (A.2) is a predictable functional of $(M_t)_{t\geq0}$. Let us precise what it exactly means. Let $\mathcal{C}(\mathbb{R}_{\geq0}, \mathbb{R}^d)$ denote the space of continuous functions $\mathbb{R}_{\geq0} \to \mathbb{R}^d$. If $w(s)$, $s \geq 0$, denote the coordinate mappings, we set $\mathcal{B}_t = \sigma(w(s), s \leq t)$. A function $F : \mathcal{C}(\mathbb{R}_{\geq0}, \mathbb{R}^d) \to \mathbb{R}^n$, is said to be a predictable functional if it is predictable as a process defined on $\mathcal{C}(\mathbb{R}_{\geq0}, \mathbb{R}^d)$ with respect to the filtration $(\mathcal{B}_t)_{t\geq0}$. So, with this terminology, saying that $(X_t^{x_0})_{t\geq0}$ is a predictable functional of $(M_t)_{t\geq0}$ means that there exists a predictable functional F satisfying for any $t \geq 0$,

$$X_t^{x_0} = F((M_s)_{0 \leq s \leq t}).$$

We mention that in general the functional F can not be chosen continuous in the topology of uniform convergence on compact sets (see Section 2.5 of Chapter 2 on the rough paths theory for a discussion about this continuity question). The essential advantage of Stratonovitch's formulation is the following form for Itô's formula.

Proposition A.3 *Let $f : \mathbb{R}^n \to \mathbb{R}$ be a C^2 function and let $(X_t^{x_0})_{t\geq0}$ denote the solution of* **SDE** $(x_0, (V_i)_{1\leq i\leq d}, (M_t)_{t\geq0})$. *We have for $t \geq 0$,*

$$f(X_t^{x_0}) = f(x_0) + \sum_{i=1}^{d} \int_0^t (V_i f)(X_s^{x_0}) \circ dM_s^i.$$

A.8 Diffusions and partial differential equations

One of the main uses of the theory of stochastic differential equations is to construct and study diffusions. This construction can be performed when the driving semimartingale $(M_t)_{t\geq0}$ is a standard Brownian motion $(B_t)_{t\geq0}$. More precisely, solving the stochastic differential equation **SDE** $(x_0, (V_i)_{1\leq i\leq d}, (B_t)_{t\geq0})$ is a way to canonically associate, on a given probability space, a stochastic process with the second order differential operator

$$\frac{1}{2} \sum_{i=1}^{d} V_i^2.$$

Indeed, as a direct consequence of Itô's formula, we get:

Proposition A.4   *Let $x_0 \in \mathbb{R}^n$ and let $(X_t^{x_0})_{t\geq 0}$ denote the solution of* **SDE** $(x_0, (V_i)_{1\leq i\leq d}, (B_t)_{t\geq 0})$. *The process $(X_t^{x_0})_{t\geq 0}$ enjoys the strong Markov property with respect to the natural filtration of $(B_t)_{t\geq 0}$. Furthermore for any smooth function $f : \mathbb{R}^n \to \mathbb{R}$ which is compactly supported, the process*

$$\left(f(X_t^{x_0}) - \int_0^t (\mathcal{L}f)(X_s^{x_0})ds \right)_{t\geq 0}$$

is a martingale with respect to the natural filtration of $(B_t)_{t\geq 0}$, where

$$\mathcal{L} = \frac{1}{2}\sum_{i=1}^d V_i^2$$

is the so-called infinitesimal generator of $(X_t^{x_0})_{t\geq 0}$.

As a consequence of this proposition, the transition function associated with the Markov process $(X_t^{x_0})_{t\geq 0}$ is the semigroup generated by $\mathcal{L}$, that is

$$P_t = e^{t\mathcal{L}}.$$

Therefore, for every $x_0 \in \mathbb{R}^n$ and every smooth function $f : \mathbb{R}^n \to \mathbb{R}$ which is compactly supported,

$$\mathbb{E}\left(f(X_t^{x_0}) \right) = \left(e^{t\mathcal{L}} f \right)(x_0).$$

It is possible to extend the last formula by adding to $\mathcal{L}$ a potential: this is the famous Feynman-Kac formula.

Proposition A.5   *Let $V : \mathbb{R}^n \to \mathbb{R}$ be a bounded and continuous function and let $f : \mathbb{R}^n \to \mathbb{R}$ be a smooth function which is compactly supported, then for every $x_0 \in \mathbb{R}^n$,*

$$\mathbb{E}\left(f(X_t^{x_0})e^{-\int_0^t V(X_s^{x_0})ds} \right) = \left(e^{t(\mathcal{L}-V)} f \right)(x_0),$$

where $(X_t^{x_0})_{t\geq 0}$ is the solution of **SDE** $(x_0, (V_i)_{1\leq i\leq d}, (B_t)_{t\geq 0})$.

A.9   Stochastic flows

As in the theory of ordinary differential equations, in the theory of stochastic differential equations, this is fruitful to look at the solution of **SDE** $(x_0, (V_i)_{1\leq i\leq d}, (M_t)_{t\geq 0})$ as a function of the initial condition x_0.

Theorem A.5 *Let* $(M_t)_{t\geq 0}$ *be a continuous d-dimensional $\mathcal{F}$-adapted semimartingale. On* $(\Omega, (\mathcal{F}_t)_{t\geq 0}, \mathbb{P})$*, there exists a unique jointly measurable continuous random field* $(\Phi_t(x))_{t\geq 0, x\in\mathbb{R}^n}$ *such that:*

(1) Almost surely, the map $\Phi_t : \mathbb{R}^n \to \mathbb{R}^n$ *is a diffeomorphism for any* $t \geq 0$*;*
(2) For $t \geq 0$ *and* $x \in \mathbb{R}^n$*,*

$$\Phi_t(x) = x + \sum_{i=1}^{d} \int_0^t V_i(\Phi_s(x)) \circ dM_s^i.$$

We call $(\Phi_t)_{t\geq 0}$ *the stochastic flow associated with the above stochastic differential equation.*

There is a nice Itô's formula for the action of stochastic flows on smooth tensor fields (we refer to Appendix B for the definitions of the tensor fields and of the Lie derivative).

Proposition A.6 *Let K be a smooth tensor field on $\mathbb{R}^n$ and let $(\Phi_t)_{t\geq 0}$ denote the stochastic flow associated with* $\mathbf{SDE}\,(x_0, (V_i)_{1\leq i\leq d}, (M_t)_{t\geq 0})$*. We have for $t \geq 0$ and $x \in \mathbb{R}^n$,*

$$(\Phi_t^* K)(x) = K(x) + \sum_{i=1}^{d} \int_0^t (\Phi_s^* \mathcal{L}_{V_i} K)(x) \circ dM_s^i.$$

A.10   Malliavin calculus

In this section, we introduce the basic tools of Malliavin calculus which are used in the proof of Hörmander's theorem. For further details, we refer to [Nualart (1995)].

Let us consider the Wiener space of continuous paths:

$$\mathbb{W}^{\otimes d} = \left(\mathcal{C}([0,1], \mathbb{R}^d), (\mathcal{B}_t)_{0\leq t\leq 1}, \mathcal{B}_1, \mathbb{P} \right)$$

where:

(1) $\mathcal{C}([0,1], \mathbb{R}^d)$ is the space of continuous functions $[0,1] \to \mathbb{R}^d$;
(2) $(B_t)_{t\geq 0}$ is the coordinate process defined by $B_t(f) = f(t)$, $f \in \mathcal{C}([0,1], \mathbb{R}^d)$;
(3) $\mathbb{P}$ is the Wiener measure on $[0,1]$, that is the law of a d-dimensional standard Brownian motion indexed by the time interval $[0,1]$;
(4) $(\mathcal{B}_t)_{0\leq t\leq 1}$ is the ($\mathbb{P}$-completed) natural filtration of $(B_t)_{0\leq t\leq 1}$.

A $\mathcal{B}_1$ measurable real valued random variable F is said to be cylindric if it can be written

$$F = f\left(\int_0^1 h_s^1 dB_s, ..., \int_0^1 h_s^n dB_s\right)$$

where $h^i \in \mathbf{L}^2([0,1], \mathbb{R}^d)$ and $f : \mathbb{R}^n \to \mathbb{R}$ is a C^∞ bounded function. The set of cylindric random variables is denoted $\mathcal{S}$.

The derivative of $F \in \mathcal{S}$ is the $\mathbb{R}^d$ valued stochastic process $(\mathbf{D}_t F)_{0 \le t \le 1}$ given by

$$\mathbf{D}_t F = \sum_{i=1}^n h^i(t) \frac{\partial f}{\partial x_i}\left(\int_0^1 h_s^1 dB_s, ..., \int_0^1 h_s^n dB_s\right).$$

More generally, we can introduce iterated derivatives. If $F \in \mathcal{S}$, we set

$$\mathbf{D}_{t_1,...,t_k}^k F = \mathbf{D}_{t_1}...\mathbf{D}_{t_k} F.$$

We can consider $\mathbf{D}^k F$ as an element of $\mathbf{L}^2\left(\mathcal{C}([0,1], \mathbb{R}^d), \mathbf{L}^2([0,1]^k, \mathbb{R}^d)\right)$; namely, $\mathbf{D}^k F$ is a random process indexed by $[0,1]^k$. For any $p \ge 1$, the operator $\mathbf{D}^k$ is closable from $\mathcal{S}$ into $\mathbf{L}^p\left(\mathcal{C}([0,1], \mathbb{R}^d), \mathbf{L}^2([0,1]^k, \mathbb{R}^d)\right)$.

We denote $\mathbb{D}^{k,p}$ the closure of the class of cylindric random variables with respect to the norm

$$\|F\|_{k,p} = \left(\mathbb{E}\left(F^p\right) + \sum_{j=1}^k \mathbb{E}\left(\|\mathbf{D}^j F\|^p_{\mathbf{L}^2([0,1]^j, \mathbb{R}^d)}\right)\right)^{\frac{1}{p}},$$

and

$$\mathbb{D}^\infty = \bigcap_{p \ge 1} \bigcap_{k \ge 1} \mathbb{D}^{k,p}.$$

We have the following key result which makes Malliavin calculus so useful when one want to study the existence of densities for random variables.

Theorem A.6 *Let $F = (F_1, ..., F_n)$ be a $\mathcal{B}_1$ measurable random vector such that:*

(1) for every $i = 1, ..., n$, $F_i \in \mathbb{D}^\infty$;
(2) the matrix

$$\Gamma = \left(\int_0^1 \langle \mathbf{D}_s F^i, \mathbf{D}_s F^j \rangle_{\mathbb{R}^d} ds\right)_{1 \le i,j \le n}$$

is invertible.

Then F has a density with respect to the Lebesgue measure. If moreover, for every $p > 1$,

$$\mathbb{E}\left(\frac{1}{|\det \Gamma|^p}\right) < +\infty,$$

then this density is smooth.

Remark A.1 *The matrix Γ is often called the Malliavin matrix of the random vector F.*

Of course, this result is only useful if the class $\mathbb{D}^\infty$ is big enough to contain interesting random variables.

Theorem A.7 *Let $(X_t^x)_{t \geq 0}$ denote the solution of the stochastic differential equation*

$$X_t^x = x + \sum_{i=1}^{d} \int_0^t V_i(X_s^{x_0}) \circ dB_s^i. \tag{A.3}$$

Then, for every $i = 1, ..., n$, $X_1^i \in \mathbb{D}^\infty$. Moreover,

$$\mathbf{D}_t^j X_1 = \mathbf{J}_{0 \to 1} \mathbf{J}_{0 \to t}^{-1} V_j(X_t), \quad j = 1, ..., d, \quad 0 \leq t \leq 1,$$

where $(\mathbf{J}_{0 \to t})_{t \geq 0}$ is the first variation process defined by

$$\mathbf{J}_{0 \to t} = \frac{\partial X_t^x}{\partial x},$$

and where $\mathbf{D}_t^j X_1^i$ is the j-th component of $\mathbf{D}_t X_1^i$.

A.11 Stochastic calculus on manifolds

Let $\mathbb{M}$ be a smooth connected manifold. A (continuous) $\mathbb{M}$-valued process $(X_t)_{t \geq 0}$ is said to be a semimartingale if for every smooth function $f : \mathbb{M} \to \mathbb{R}$ the process $(f(X_t))_{t \geq 0}$ is a real semimartingale. It is interesting to note that to define the notion of semimartingale on a manifold, no additional structure than the differentiability structure is required. But to define the notion of martingale on a manifold, we have to endow the manifold with an affine connection. On this point we shall not go into details since the notion of martingales on a manifold is not needed in this book, but for further details we refer to [Emery (1989)] and [Emery (2000)].

On the manifold $\mathbb{M}$, it is possible to give a sense to stochastic differential equations of the type

$$X_t^{x_0} = x_0 + \sum_{i=1}^{d} \int_0^t V_i(X_s^{x_0}) \circ dM_s^i, \quad t \geq 0 \tag{A.4}$$

where:

(1) $x_0 \in \mathbb{M}$;
(2) $V_1, ..., V_d$ are C^∞ bounded vector fields on $\mathbb{M}$;
(3) $\circ$ denotes Stratonovitch integration;
(4) $(M_t^1, ..., M_t^d)_{0 \leq t \leq T}$ is a $\mathbb{R}^d$-valued continuous semimartingale.

Indeed, we shall say that a $\mathbb{M}$-valued process is solution of (A.4) if for every smooth function $f : \mathbb{M} \to \mathbb{R}$,

$$f(X_t^{x_0}) = f(x_0) + \sum_{i=1}^{d} \int_0^t (V_i f)(X_s^{x_0}) \circ dM_s^i, \quad t \geq 0.$$

In this setting, many results given in Sections A7, A8 and A9 remain true. In particular, when the driving semimartingale $(M_t)_{t \geq 0}$ is a linear Brownian motion, then the solution of equation (A.4) is a Markov process with infinitesimal generator

$$\frac{1}{2} \left(\sum_{i=1}^{d} V_i^2 \right).$$

Let us now assume that the manifold $\mathbb{M}$ is moreover endowed with a Riemannian structure (see Appendix B). In that case, there is a natural and canonical second-order elliptic operator defined on $\mathbb{M}$: The Laplace-Beltrami operator $\Delta_\mathbb{M}$.

A continuous $\mathbb{M}$-valued process $(B_t)_{t \geq 0}$ is said to be a Brownian motion on $\mathbb{M}$ if it is a Markov process with generator $\frac{1}{2} \Delta_\mathbb{M}$. It is equivalent to ask that for every C^∞ bounded function $\mathbb{M} \to \mathbb{R}$, the process

$$\left(f(B_t) - \frac{1}{2} \int_0^t (\Delta_\mathbb{M} f)(B_s) \, ds \right)_{t \geq 0}$$

is a martingale.

It may happen that a Brownian motion on a manifold explodes, meaning that the best we can do is to find a process $(B_t)_{t \geq 0}$ together with a stopping time τ such that for every C^∞ bounded function $\mathbb{M} \to \mathbb{R}$, the process

$$\left(f\left(B_{\inf(t,\tau)}\right) - \frac{1}{2} \int_0^{\inf(t,\tau)} \left(\Delta_{\mathbb{M}} f\right)\left(B_s\right) ds \right)_{t \geq 0}$$

is a martingale. Such phenomenon can not occur if the manifold $\mathbb{M}$ is compact.

Observe that in general the operator $\Delta_{\mathbb{M}}$ can not globally be written under the form

$$V_0 + \sum_{i=1}^{d} V_i^2,$$

where $V_0, V_1, ..., V_d$ are smooth vector fields on $\mathbb{M}$. Therefore, in full generality, we can not construct Brownian motions on a Riemannian manifold by solving stochastic differential equations on the manifold. An intrinsic way to construct Brownian motions on a manifold by solving a stochastic differential equation can be done by working in the orthonormal frame bundle of the manifold: this the Eels-Elworthy-Malliavin construction. This construction is widely explained in the Section 3.4. of Chapter 3.

time τ such that for every C^2 bounded function $M \to \mathbb{R}$, the process

$$\left(f(B_{t \wedge \tau}) - \int_0^{t \wedge \tau} \frac{1}{2} (\Delta_M f)(B_s) ds \right)_{t \geq 0}$$

is a martingale... local... can not occur if the manifold M is compact.

Observe that in general the operator Δ_M can not globally be written under the form

$$N + \sum_{i=1}^{n} V_i^2$$

where $N, V_1, \ldots, V_n$ are smooth vector fields on M. Therefore, in full generality we can not construct Brownian motions on a Riemannian manifold by solving stochastic differential equations on the manifold. An intrinsic way to construct Brownian motions on a manifold by solving a stochastic differential equation can be done by working in the orthonormal frame bundle of the manifold: this is the Eels-Elworthy-Malliavin construction. This construction is widely explained in the Section 4.x. of Chapter 3.

Appendix B

Vector Fields, Lie Groups and Lie Algebras

B.1 Vector fields and exponential mapping

Let $\mathcal{O} \subset \mathbb{R}^n$ be a non empty open set. A smooth vector field V on $\mathcal{O}$ is simply a smooth map

$$V : \mathcal{O} \to \mathbb{R}^n$$
$$x \to (v_1(x), ..., v_n(x)).$$

It is a basic result in the theory of ordinary differential equations that if $K \subset \mathcal{O}$ is compact, there exist $\varepsilon > 0$ and a smooth mapping

$$\Phi : (-\varepsilon, \varepsilon) \times K \to \mathcal{O},$$

such that for $x \in K$ and $-\varepsilon < t < \varepsilon$,

$$\frac{\partial \Phi}{\partial t}(t, x) = X(\Phi(t, x)), \ \Phi(0, x) = x.$$

Furthermore, if $y : (-\eta, \eta) \to \mathbb{R}^n$ is a C^1 path such that for $-\eta < t < \eta$, $y'(t) = X(y(t))$, then $y(t) = \Phi(t, y(0))$ for $-\min(\eta, \varepsilon) < t < \min(\eta, \varepsilon)$. From this characterization of Φ it is easily seen that for $x \in K$ and $t_1, t_2 \in \mathbb{R}$ such that $|t_1| + |t_2| < \varepsilon$,

$$\Phi(t_1, \Phi(t_2, x)) = \Phi(t_1 + t_2, x).$$

Because of this last property, the solution mapping $t \to \Phi(t, x)$ is called the exponential mapping, and we denote $\Phi(t, x) = e^{tV}(x)$. It always exists if $|t|$ is sufficiently small. If e^{tV} can be defined for any $t \in \mathbb{R}$, then the vector field is said to be complete. For instance if $\mathcal{O} = \mathbb{R}^n$ and if V is C^∞-bounded then the vector field V is complete.

The vector field V defines a differential operator acting on the smooth functions $f : \mathcal{O} \to \mathbb{R}$ as follows:

$$(\mathcal{L}_V f)(x) = \left(\frac{d}{dt} \right)_{t=0} f(e^{tV}(x)), \ x \in \mathcal{O}.$$

We also use the following notation

$$(\mathcal{L}_V f)(x) = (Vf)(x),$$

that is, we apply V to f as a first-order differential operator. With this notation, observe that

$$V(x) = \sum_{i=1}^{n} v_i(x) \frac{\partial}{\partial x_i}. \tag{B.1}$$

We note that V is a derivation, that is a map on $C^\infty(\mathcal{O}, \mathbb{R})$, linear over $\mathbb{R}$, satisfying for $f, g \in C^\infty(\mathcal{O}, \mathbb{R})$,

$$V(fg) = (Vf)g + f(Vg).$$

An interesting result is that, conversely, any derivation on $C^\infty(\mathcal{O}, \mathbb{R})$ is a vector field, i.e. has the form (B.1). If V' is another smooth vector field on $\mathcal{O}$, then it is easily seen that the operator $VV' - V'V$ is a derivation. It therefore defines a smooth vector field on $\mathcal{O}$ which is called the Lie bracket of V and V' and denoted $[V, V']$. A straightforward computation shows that for $x \in \mathcal{O}$,

$$[V, V'](x) = \sum_{i=1}^{n} \left(\sum_{j=1}^{n} v_j(x) \frac{\partial v'_i}{\partial x_j}(x) - v'_j(x) \frac{\partial v_i}{\partial x_j}(x) \right) \frac{\partial}{\partial x_i}.$$

Observe that the Lie bracket satisfies obviously $[V, V'] = -[V', V]$ and the so-called Jacobi identity, that is:

$$[V, [V', V'']] + [V', [V'', V]] + [V'', [V, V']] = 0.$$

Let us now give a geometric meaning of the bracket. If $\phi : \mathcal{O}' \to \mathcal{O}$ is a diffeomorphism between two open domains in $\mathbb{R}^n$, the pull-back $\phi^* V$ of the vector field V by the map ϕ is the vector field on $\mathcal{O}'$ defined by the chain rule,

$$\phi^* V(x) = (d\phi^{-1})_{\phi(x)} \left(V(\phi(x)) \right), \ x \in \mathcal{O}'.$$

In particular, if V' is a smooth vector field on $\mathcal{O}$, $(e^{tV})^*V'$ is defined on most of $\mathcal{O}$, for $|t|$ small, and we can define the Lie derivative,

$$\mathcal{L}_V V' = \left(\frac{d}{dt}\right)_{t=0} (e^{tV})^*V'$$

as a vector field on $\mathcal{O}$. It actually turns out that we have

$$\mathcal{L}_V V' = [V, V'].$$

If the vector fields V and V' commute, that is if $[V, V'] = 0$ then the flows they generate locally commute; in other words, for any $x \in \mathcal{O}$, there exists a $\delta > 0$ such that for $|t_1| + |t_2| < \delta$,

$$e^{t_1 V} e^{t_2 V'}(x) = e^{t_2 V'} e^{t_1 V}(x).$$

Another essential point on the bracket is that one can *flow* in the direction of the commutator $[V, V']$ by a succession of flows along V and V'. More precisely, we have

$$e^{-tV} e^{-tV'} e^{tV} e^{tV'} = e^{t^2[V,V']} + O(t^3).$$

B.2   Lie derivative of tensor fields along vector fields

A smooth tensor field K of type (r, s), $r, s \in \mathbb{N}$, on $\mathbb{R}^n$ is a smooth map

$$K : \mathcal{O} \to (\mathbb{R}^n)^{\otimes r} \otimes (\mathbb{R}^{n*})^{\otimes s},$$

where $\mathbb{R}^{n*}$ denotes the dual space of $\mathbb{R}^n$. By convention a $(0,0)$ tensor is a smooth function $f : \mathcal{O} \to \mathbb{R}$. Any (r, s) smooth tensor field K can therefore be written

$$K = V_1 \otimes \cdots \otimes V_r \otimes \alpha_1 \otimes \cdots \otimes \alpha_s,$$

where the V_i's are smooth vector fields and the α_i's smooth one-forms, i.e. applications $\mathcal{O} \to \mathbb{R}^{n*}$. Let $\phi : \mathcal{O}' \to \mathcal{O}$ denote a diffeomorphism between two open domains in $\mathbb{R}^n$, the pull-back of a smooth one-form $\alpha : \mathcal{O} \to \mathbb{R}^{n*}$ by the map ϕ is the smooth one-form on $\mathcal{O}'$ defined by,

$$\phi^*\alpha(x) = \alpha_{\phi(x)}(d\phi(x)), \ x \in \mathcal{O}'.$$

The pull-back by ϕ can now be defined on any (r, s) smooth tensor field K in the following way:

(1) If K is of type $(0,0)$, i.e. if K is a smooth function $f : \mathcal{O} \to \mathbb{R}$, then

$$\phi^* f(x) = (f \circ \phi)(x), \ x \in \mathcal{O}';$$

(2) If

$$K = V_1 \otimes \cdots \otimes V_r \otimes \alpha_1 \otimes \cdots \otimes \alpha_s,$$

then

$$\phi^* K = \phi^* V_1 \otimes \cdots \otimes \phi^* V_r \otimes \phi^* \alpha_1 \otimes \cdots \otimes \phi^* \alpha_s.$$

Thanks to this action we are now able to define the notion of Lie derivative along a given vector field. Let V denote a smooth vector field on $\mathcal{O}$ and let K denote a smooth (r, s) tensor field also defined on $\mathcal{O}$. We define the Lie derivative of K along V as the smooth (r, s) tensor field on $\mathcal{O}$ given by

$$\mathcal{L}_V K = \left(\frac{d}{dt} \right)_{t=0} (e^{tV})^* K.$$

The Lie derivative along a vector field V enjoys the following properties:

(1) If f is a smooth function $\mathcal{O} \to \mathbb{R}$, then

$$\mathcal{L}_V f = V f;$$

(2) If K is a smooth (r, s) tensor field and V' a smooth vector field

$$\mathcal{L}_{[V,V']} K = [\mathcal{L}_V, \mathcal{L}_{V'}] K = (\mathcal{L}_V \mathcal{L}_{V'} - \mathcal{L}_{V'} \mathcal{L}_V) K.$$

B.3   Exterior forms and exterior derivative

For $k \geq 0$, denote $\wedge^k \mathbb{R}^n$ the space of skew symmetric k multilinear maps $(\mathbb{R}^n)^p \to \mathbb{R}$. For $\alpha \in \wedge^k \mathbb{R}^n, \beta \in \wedge^l \mathbb{R}^n$, we define $\alpha \wedge \beta \in \wedge^{k+l} \mathbb{R}^n$ in the following way: For any $(u_1, ..., u_{k+l}) \in (\mathbb{R}^n)^{k+l}$,

$$(\alpha \wedge \beta)(u_1, ..., u_{k+l}) = \sum \mathrm{sign}(\sigma) \alpha(u_{\sigma(1)}, ..., u_{\sigma(k)}) \beta(v_{\sigma(k+1)}, ..., v_{\sigma(k+l)}),$$

where the sum is taken over all *shuffles*; that is, permutations σ of $\{1, ..., k+l\}$ such that $\sigma(1) < ... < \sigma(k)$ and $\sigma(k+1) < ... < \sigma(k+l)$. Besides the bilinearity, the basic properties of the wedge product $\wedge$ are the following. For $\alpha \in \wedge^k \mathbb{R}^n, \beta \in \wedge^l \mathbb{R}^n, \gamma \in \wedge^m \mathbb{R}^n$:

(1) $\alpha \wedge \beta = (-1)^{kl} \beta \wedge \alpha$;
(2) $\alpha \wedge (\beta \wedge \gamma) = (\alpha \wedge \beta) \wedge \gamma$.

If $(\alpha^1, ..., \alpha^n)$ denotes the canonical basis of the dual space of $\mathbb{R}^n$, then it is easily seen that the family

$$\{\alpha^{i_1} \wedge ... \wedge \alpha^{i_k}, 1 \leq i_1 < ... < i_k \leq n\}$$

is a basis for $\wedge^k \mathbb{R}^n$. It follows that for $k > n$, $\wedge^k \mathbb{R}^n = \{0\}$, while for $0 < k \leq n$, $\wedge^k \mathbb{R}^n$ is a vector space with dimension $\binom{n}{k}$. The direct sum of the spaces $\wedge^k \mathbb{R}^n$, $k \geq 0$ together with its structure as a real vector space and multiplication induced by $\wedge$, is called the exterior algebra of $\mathbb{R}^n$. It is a vector space with dimension 2^n, which is denoted $\wedge \mathbb{R}^n$.

An exterior differential k-form on $\mathbb{R}^n$ is a smooth mapping $\alpha : \mathbb{R}^n \to \wedge^k \mathbb{R}^n$. The set of differential k-forms shall be denoted $\Omega^k(\mathbb{R}^n)$. The following fact is fundamental. There is a unique family of mappings $\mathbf{d}^k : \Omega^k(\mathbb{R}^n) \to \Omega^{k+1}(\mathbb{R}^n)$ $(k = 0, ..., n)$, which we merely denote by $\mathbf{d}$, called the exterior derivative on $\mathbb{R}^n$ such that:

(1) $\mathbf{d}$ is a $\wedge$ antiderivation. That is, $\mathbf{d}$ is $\mathbb{R}$ linear and for $\alpha \in \Omega^k(\mathbb{R}^n), \beta \in \Omega^l(\mathbb{R}^n)$,

$$\mathbf{d}(\alpha \wedge \beta) = \mathbf{d}\alpha \wedge \beta + (-1)^k \alpha \wedge \mathbf{d}\beta;$$

(2) If $f : \mathbb{R}^n \to \mathbb{R}$ is a smooth map, then $\mathbf{d}f$ is the differential of f;
(3) $\mathbf{d} \circ \mathbf{d} = 0$ (that is, $\mathbf{d}^{k+1} \circ \mathbf{d}^k = 0$).

The link between the exterior derivative and the Lie bracket of vector fields is given by the following formula which is due to E. Cartan: For any differential one-form α and any smooth vector fields U and V,

$$\mathbf{d}\alpha(U, V) = (\mathcal{L}_U \alpha)(V) - (\mathcal{L}_V \alpha)(U) - \alpha([U, V]),$$

where $\mathcal{L}$ denotes the Lie derivative.

B.4 Lie groups and Lie algebras

A Lie group $\mathbb{G}$ is a group that is also an analytic manifold, such that the group operations $\mathbb{G} \times \mathbb{G} \to \mathbb{G}$ and $\mathbb{G} \to \mathbb{G}$ given by $(g, h) \to gh$ and $g \to g^{-1}$ are analytic maps. Let $\mathbf{e}$ denote the identity element of $\mathbb{G}$. For each $g \in \mathbb{G}$ we have left and right translations $\mathbf{L}_g$ and $\mathbf{R}_g$, diffeomorphisms on $\mathbb{G}$ defined by $\mathbf{L}_g(h) = gh$ and $\mathbf{R}_g(h) = hg$. A smooth vector field V on

$\mathbb{G}$ is said to be left-invariant if for any $g \in \mathbb{G}$,

$$\mathbf{L}_g^* V = V.$$

The set of left-invariant vector fields on $\mathbb{G}$ is a linear space called the Lie algebra of $\mathbb{G}$ and is denoted $\mathfrak{g}$. The evaluation map $V \to V(\mathbf{e})$ provides a linear isomorphism between $\mathfrak{g}$ and the tangent space to $\mathbb{G}$ at the identity. Therefore $\mathfrak{g}$ is a finite dimensional vector space whose dimension is the dimension of $\mathbb{G}$. An important point is that $\mathfrak{g}$ is closed under the bracket operation. Namely, for $V_1, V_2 \in \mathfrak{g}$, $[V_1, V_2] \in \mathfrak{g}$. There is a natural map from $\mathfrak{g}$ to $\mathbb{G}$, the exponential map, defined as follows. It is possible to show that a left-invariant vector field V defines a global flow on $\mathbb{G}$, that is e^{tV} is defined for every $t \in \mathbb{R}$. In particular, it is defined for $t = 1$. The map $\mathfrak{g} \to \mathbb{G}$, $V \to e^V(\mathbf{e})$, is called the exponential map (by a slight abuse of notation, we shall still denote $e^V = e^V(\mathbf{e})$). The exponential map is a diffeomorphism from a neighborhood of 0 in $\mathfrak{g}$ onto a neighborhood of $\mathbf{e}$ in $\mathbb{G}$, and its differential map at 0 is the identity. Nevertheless, let us mention that, in general, the exponential map is neither a global diffeomorphism, nor a local homomorphism, nor even a surjection. It is easily seen from the properties of the exponential of vector fields that for $V \in \mathfrak{g}$, $s, t \in \mathbb{R}$,

$$e^{sV} e^{tV} = e^{(s+t)V}.$$

More generally, if $V_1, V_2 \in \mathfrak{g}$ commute, that is $[V_1, V_2] = 0$, then

$$e^{V_1} e^{V_2} = e^{V_2} e^{V_1}.$$

It is of course possible to define the notion of Lie algebra independently of the notion of Lie group. Namely, an abstract Lie algebra is a vector space $\mathfrak{L}$, equipped with a bilinear map $[\cdot, \cdot] : \mathfrak{L} \times \mathfrak{L} \to \mathfrak{L}$ such that:

(1) $[X, Y] = -[Y, X]$ for $X, Y \in \mathfrak{L}$;
(2) $[X, [Y, Z]] + [Y, [Z, X]] + [Z, [X, Y]] = 0$ for $X, Y, Z \in \mathfrak{L}$ (*Jacobi Identity*).

If $\mathfrak{L}$ is an abstract Lie algebra, the universal envelopping algebra $\mathcal{A}(\mathfrak{L})$ is defined as

$$\mathcal{A}(\mathfrak{L}) = \mathfrak{L}_{\mathbb{C}}^{\otimes} / \mathcal{J},$$

where,

(1) $\mathfrak{L}_{\mathbb{C}}$ is the complexification of $\mathfrak{g}$,
(2) $\mathfrak{L}_{\mathbb{C}}^{\otimes} = \oplus_{k=0}^{\infty} \mathfrak{L}_{\mathbb{C}}^{\otimes k}$,

(3) $\mathcal{J}$ is the two-sided ideal generated by

$$\{X \otimes Y - Y \otimes X - [X,Y], X, Y \in \mathfrak{L}_{\mathbb{C}}\}.$$

In the case where $\mathfrak{L}$ is the Lie algebra of a Lie group $\mathbb{G}$, $\mathcal{A}(\mathfrak{L})$ can be identified with the set of all left-invariant operators on $\mathbb{G}$. Besides vector fields there are many interesting left-invariant differential operators on $\mathbb{G}$. For instance, up to a constant multiple, there is a unique left-invariant measure on $\mathbb{G}$ which is called a (left) Haar measure. In many but not all cases left Haar measure is also right Haar measure; In that case $\mathbb{G}$ is said to be unimodular. For instance all compact Lie groups are unimodular. Besides the Carnot groups, studied extensively in this book, there are some particular Lie groups of matrices that we want to mention.

Example B.1 In the following examples we indicate the Lie groups and the corresponding Lie algebras. The exponential map is simply the usual exponential of matrices and the Lie bracket is given by the commutator $[A,B] = AB - BA$. Let $n \in \mathbb{N}$, $n > 1$, $\mathbb{K} = \mathbb{R}$ or $\mathbb{C}$, and $\mathcal{M}_n(\mathbb{K})$ the set of $n \times n$ matrices with entries in $\mathbb{K}$.

(1)

$$\mathbf{GL}_n(\mathbb{K}) = \{A \in \mathcal{M}_n(\mathbb{K}), \det A \neq 0\}, \ \mathfrak{gl}_n(\mathbb{K}) = \mathcal{M}_n(\mathbb{K}).$$

(2)

$$\mathbf{SL}_n(\mathbb{K}) = \{A \in \mathcal{M}_n(\mathbb{K}), \det A = 1\}, \ \mathfrak{sl}_n(\mathbb{K}) = \{A \in \mathcal{M}_n(\mathbb{K}), \mathrm{Tr} A = 0\}.$$

(3)

$$\mathbf{O}_n(\mathbb{R}) = \{A \in \mathbf{GL}_n(\mathbb{R}), A^t = A^{-1}\}, \ \mathfrak{o}_n(\mathbb{R}) = \{A \in \mathcal{M}_n(\mathbb{R}), A^t = -A\}.$$

(4)

$$\mathbf{U}_n(\mathbb{C}) = \{A \in \mathbf{GL}_n(\mathbb{C}), A^* = A^{-1}\}, \ \mathfrak{u}_n(\mathbb{C}) = \{A \in \mathcal{M}_n(\mathbb{C}), A^* = -A\}.$$

B.5 The Baker-Campbell-Hausdorff formula

The Baker-Campbell-Hausdorff formula shows, and this is a priori a non obvious fact, that the multiplication law in a Lie group is determined uniquely and very explicitly by the Lie algebra structure, at least in a neighborhood of the identity. Let $\mathbb{G}$ be a Lie group and let $\mathfrak{g}$ be its Lie algebra. We have seen that if $X, Y \in \mathfrak{g}$ satisfy $[X,Y] = 0$ then $e^X e^Y = e^Y e^X$. It is actually easy to check that under this commutation assumption, we have $e^X e^Y = e^{X+Y}$. This formula is not true for arbitrary $X, Y \in \mathfrak{g}$. Nevertheless, the question arises naturally whether one can obtain an explicit formula for $e^X e^Y$. This is the content of the Baker-Campbell-Hausdorff formula which reads

$$e^X e^Y = e^{\mathcal{P}(X,Y)},$$

where X and Y are elements of $\mathfrak{g}$ in a sufficiently small neighborhood U of 0, and where the map $\mathcal{P} : U \times U \to \mathfrak{g}$ has a universal form which is independent of $\mathbb{G}$. Let us precise the form of $\mathcal{P}$ (the formula we give comes from [Dynkin (1947)], but referred to as the Specht-Wever theorem in [Jacobson (1962)]). For this, we introduce some notations. For $X \in \mathfrak{g}$, let $\mathbf{ad}X$ denote the linear endomorphism $\mathfrak{g} \to \mathfrak{g}$ given by $(\mathbf{ad}X)(Y) = [X, Y]$, $Y \in \mathfrak{g}$. We have,

$$\mathcal{P}(X,Y) = \sum_{k=1}^{+\infty} \frac{(-1)^{k-1}}{k} \sum \frac{(\mathbf{ad}Y)^{q_k}(\mathbf{ad}X)^{p_k} \cdots (\mathbf{ad}Y)^{q_1}(\mathbf{ad}X)^{p_1}}{\left(\sum_{l=1}^{k}(p_l + q_l)\right)\left(\prod_{l=1}^{k} p_l! q_l!\right)}, \quad \text{(B.2)}$$

where the inner sum is over the set of nonnegative integers (p_i, q_i) such that

$$p_i + q_i > 0,$$

(of course either $p_1 = 1$ or $p_1 = 0, q_1 = 1$), and we used the convention $(\mathbf{ad}X)^1 = X$. We have a simple generalization of (B.2) for more than two exponentials:

$$e^{X_1} \cdots e^{X_n} = e^{\mathcal{P}(X_1, \cdots, X_n)},$$

where $\mathcal{P}(X_1, \cdots, X_n)$ is given by

$$\sum_{k=1}^{+\infty} \frac{(-1)^{k-1}}{k} \sum \frac{(\mathbf{ad}X_n)^{p_{k,n}} \cdots (\mathbf{ad}X_1)^{p_{k,1}} \cdots (\mathbf{ad}X_n)^{p_{1,n}} \cdots (\mathbf{ad}X_1)^{p_{1,1}}}{\left(\sum_{l,m=1}^{k} p_{l,m}\right)\left(\prod_{l,m=1}^{k} p_{l,m}!\right)},$$

the inner sum being taken over all nonnegative integers such that

$$\sum_{m=1}^{n} p_{l,m} > 0, \; l = 1, \cdots, k.$$

Actually, the universality of $\mathcal{P}$ actually stems from a purely algebraic identity between formal series. Indeed, let us denote by $\mathbb{R}[[X_1, ..., X_d]]$ the non-commutative algebra of formal series with d indeterminates. The exponential of $Y \in \mathbb{R}[[X_1, ..., X_d]]$ is defined by

$$e^Y = \sum_{k=0}^{+\infty} \frac{Y^k}{k!}.$$

Now, define the bracket between two elements Y and Z of $\mathbb{R}[[X_1, ..., X_d]]$ by

$$[Y, Z] = YZ - ZY,$$

and denote by **ad** the map defined by $(\mathbf{ad}Y)Z = [Y, Z]$. In this context, the Baker-Campbell-Hausdorff formula reads

$$e^Y e^Z = e^{\mathcal{P}(Y,Z)},$$

where

$$\mathcal{P}(Y, Z) = \sum_{k=1}^{+\infty} \frac{(-1)^{k-1}}{k} \sum \frac{(\mathbf{ad}Z)^{q_k}(\mathbf{ad}Y)^{p_k} \cdots (\mathbf{ad}Z)^{q_1}(\mathbf{ad}Y)^{p_1}}{\left(\sum_{l=1}^{k}(p_l + q_l)\right)\left(\prod_{l=1}^{k} p_l! q_l!\right)}, \quad (B.3)$$

and, as before, the inner sum is over the set of nonnegative integers (p_i, q_i) such that

$$p_i + q_i > 0.$$

Observe now that since

$$e^Y e^Z = 1 + \sum_{p+q>0} \frac{Y^p Z^q}{p! q!},$$

and

$$\ln\left(e^Y e^Z\right) = \sum_{m=1}^{+\infty} \frac{(-1)^{m-1}}{m} \left(e^Y e^Z - 1\right)^m,$$

the formula (B.3) can also be written as

$$\sum_{m=1}^{+\infty} \frac{(-1)^{m-1}}{m} \left(\sum_{p+q>0} \frac{Y^p Z^q}{p! q!}\right)^m = \mathcal{P}(Y, Z).$$

More generally, in the same way, we have for $Y_1, ..., Y_n \in \mathbb{R}[[X_1, ..., X_d]]$,

$$\sum_{m=1}^{+\infty} \frac{(-1)^{m-1}}{m} \left(\sum_{p_1+\cdots+p_n>0} \frac{Y_1^{p_1} \cdots Y_n^{p_n}}{p_1! \cdots p_n!}\right)^m = \mathcal{P}(Y_1, \cdots, Y_n), \quad (B.4)$$

where $\mathcal{P}(Y_1, \cdots, Y_n)$ is given by

$$\sum_{k=1}^{+\infty} \frac{(-1)^{k-1}}{k} \sum \frac{(\mathrm{ad}Y_n)^{p_{k,n}} \cdots (\mathrm{ad}Y_1)^{p_{k,1}} \cdots (\mathrm{ad}Y_n)^{p_{1,n}} \cdots (\mathrm{ad}Y_1)^{p_{1,1}}}{\left(\sum_{l,m=1}^k p_{l,m}\right) \left(\prod_{l,m=1}^k p_{l,m}!\right)},$$

the inner sum being, as before, taken over all nonnegative integers such that

$$\sum_{m=1}^n p_{l,m} > 0, \ l = 1, \cdots, k.$$

This is this version of the Baker-Campbell-Hausdorff formula which is used in the proof of the Chen-Strichartz formula.

B.6   Nilpotent Lie groups

An abstract Lie algebra $\mathfrak{L}$ is said to be nilpotent if for any $X \in \mathfrak{L}$, the map $\mathrm{ad}X : Y \to [X, Y]$ is a nilpotent endomorphism of $\mathfrak{L}$. A Lie group $\mathbb{G}$ is said to be nilpotent if its Lie algebra $\mathfrak{g}$ is a nilpotent Lie algebra. If $\mathbb{G}$ is a nilpotent Lie group, then the Baker-Campbell-Hausdorff is global. Indeed, in that case, from Dynkyn's formula (B.2) there exists a Lie polynomial $\mathcal{P} : \mathfrak{g} \times \mathfrak{g} \to \mathfrak{g}$, with rational coefficients, such that for any $X, Y \in \mathfrak{g}$,

$$e^X e^Y = e^{\mathcal{P}(X,Y)}.$$

According to (B.2), the first terms of $\mathcal{P}$ are the following:

$$\mathcal{P}(X,Y) = X + Y + \tfrac{1}{2}[X,Y] + \tfrac{1}{12}[[X,Y],Y] - \tfrac{1}{12}[[X,Y],X]$$
$$- \tfrac{1}{48}[Y,[X,[X,Y]]] - \tfrac{1}{48}[X,[Y,[X,Y]]] + \cdots.$$

Moreover if $\mathbb{G}$ is simply connected then the exponential map $\mathfrak{g} \to \mathbb{G}$ is a diffeomorphism. In that case, the group law of $\mathbb{G}$ is thus fully characterized by the Lie algebra structure of $\mathfrak{g}$. It can also be shown that a nilpotent Lie group can always be seen as a group of unipotent matrices.

B.7   Free Lie algebras and Hall basis

In this section we describe the construction of a linear basis in the free Lie algebra with d generators. Let Z be the set of the d indeterminates $X_1, ..., X_d$. Let $\mathfrak{L}_0$ be an abstract Lie algebra and $i : Z \to \mathfrak{L}_0$ a mapping. The Lie algebra $\mathfrak{L}_0$ is called free over Z if for any abstract Lie algebra $\mathfrak{L}$

and any mapping $f : Z \to \mathfrak{L}$, there is a unique Lie algebra homomorphism $\tilde{f} : \mathfrak{L}_0 \to \mathfrak{L}$ such that $f = \tilde{f} \circ i$. It can be shown that there is a free Lie algebra over Z which is unique up to isomorphism. This Lie algebra is called the free Lie algebra with the d generators $X_1, ..., X_d$. It shall be denoted $\mathfrak{L}(X_1, ..., X_d)$. Denote now $\mathcal{M}(Z)$ the free monoid over Z, $l(h)$ the length of a word $l \in \mathcal{M}(Z)$ and $\mathcal{M}^i(Z)$ the set of all length i words. A Hall family over Z is an arbitrary linearly ordered subset $H \subset \mathcal{M}(Z)$ such that:

(1) If $u, v \in H$ and $l(u) < l(v)$ then $u < v$;
(2) $Z \subset H$;
(3) $H \cap \mathcal{M}^2(Z) = \{xy, x, y \in Z, x < y\}$;
(4) $H - (\mathcal{M}^2(Z) \cup Z) = \{a(bc), a, b, c \in H, b \le a < bc, b < c\}$.

Define a mapping $\theta : \mathcal{M}(Z) \to \mathfrak{L}(X_1, ..., X_d)$ as follows. Take a word from $\mathcal{M}(Z)$ and replace all round brackets with Lie brackets (e.g. $X_1 X_2$ becomes $[X_1, X_2]$, $X_1(X_1 X_2)$ becomes $[X_1, [X_1, X_2]]$, etc...). Then the Hall-Witt theorem asserts that θ maps H into a homogeneous linear basis in $\mathfrak{L}(X_1, ..., X_d)$.

B.8 Basic Riemannian geometry

Let $V_1, ..., V_n$ be C^∞ bounded vector fields on $\mathbb{R}^n$ such that for every $x \in \mathbb{R}^n$, $(V_1(x), ..., V_n(x))$ is a basis of $\mathbb{R}^n$;
Let us denote $(\theta^1, ..., \theta^n) \in \Omega^1(\mathbb{R}^n)^n$ the dual basis of $(V_1, ..., V_n)$. The first invariant associated with the system $(V_1, ..., V_n)$ is the family of scalar products $(g_x)_{x \in \mathbb{R}^n}$ on $\mathbb{R}^n$ obtained by declaring that for any $x \in \mathbb{R}^n$, the family $(V_1(x), ..., V_n(x))$ is orthonormal. The following theorem is the fundamental theorem of Riemannian geometry.

Theorem B.1 *There is a unique matrix of one-forms* $\omega = (\omega^i_j)_{1 \le i, j \le n}$ *such that:*

(1) $d\theta + \omega \wedge \theta = 0$, *that is, for any* $1 \le i \le n$, $d\theta^i + \sum_{j=1}^n \omega^i_j \wedge \theta^j = 0$;
(2) $\omega^t = -\omega$, *that is, for any* $1 \le i, j \le n$, $\omega^i_j = -\omega^j_i$.

The matrix of one-forms ω is called the connection form, and the equations

$$d\theta + \omega \wedge \theta = 0, \quad \omega^t = -\omega,$$

the first structural equations. Recall now that a connection ∇ on $\mathbb{R}^n$ is simply a convention for differentiating a vector field along another vector field. If we denote by $\mathcal{V}(\mathbb{R}^n)$ the set of smooth vector fields on $\mathbb{R}^n$, it is more precisely a map

$$\nabla : \mathcal{V}(\mathbb{R}^n) \times \mathcal{V}(\mathbb{R}^n) \to \mathcal{V}(\mathbb{R}^n)$$

such that for any $U, V, W \in \mathcal{V}(\mathbb{R}^n)$ and any smooth $f, g : \mathbb{R}^n \to \mathbb{R}$:

(1) $\nabla_{fU+gV} W = f\nabla_U W + g\nabla_V W$;
(2) $\nabla_U(V + W) = \nabla_U V + \nabla_U W$;
(3) $\nabla_U(fV) = f\nabla_U V + U(f)V$.

Since, by assumption, for any $x \in \mathbb{R}^n$, the family of vectors $(V_1(x), ..., V_n(x))$ is a basis for $\mathbb{R}^n$, it is easily seen that the vector fields $\nabla_{V_i} V_j$, $1 \le i, j \le n$ entirely characterize a connection ∇. Thus, by using our connection form ω we can generate a connection ∇ by declaring that for any $1 \le i, j \le n$,

$$\nabla_{V_i} V_j = \sum_{k=1}^{n} \omega_j^k(V_i) V_k.$$

The connection ∇ is called the Levi-Civita connection associated with the elliptic system $(V_1, ..., V_n)$. This connection enjoys the two following additional properties:

(1) It is torsion free, that is, for any smooth vector fields X, Y,

$$\nabla_U V - \nabla_V U = [U, V];$$

(2) It is metric, that is, for any smooth vector fields U, V, W,

$$U(g(V, W)) = g(\nabla_U V, W) + g(V, \nabla_U W).$$

Let us now turn to the second structural equations. The equations

$$\Omega = \mathbf{d}\omega + \omega \wedge \omega,$$

that is,

$$\Omega_j^i = \mathbf{d}\omega_j^i + \sum_{k=1}^{n} \omega_k^i \wedge \omega_j^k, \ 1 \le i, j \le n,$$

define a skew-symmetric matrix of 2-forms such that for any smooth vector fields X, Y,

$$\sum_{i=1}^{n} \Omega_j^i(X, Y) V_i = \nabla_X \nabla_Y V_j - \nabla_Y \nabla_X V_j - \nabla_{[X,Y]} V_j, \ 1 \le j \le n.$$

The $(1,3)$-tensor $\mathbf{R}$ defined by the property that for any smooth vector fields X, Y, Z,

$$\mathbf{R}(X,Y)Z = \nabla_X \nabla_Y Z - \nabla_Y \nabla_X Z - \nabla_{[X,Y]} Z$$

is called the Riemannian curvature tensor of ∇ while the matrix of two-forms Ω is called the curvature form. The Ricci curvature $\mathbf{Ric}$ is a trace of the Riemannian curvature, it is the $(0,2)$ tensor defined by

$$\mathbf{Ric}(X,Y) = \sum_{i=1}^{n} g\left(\mathbf{R}(X,V_i)V_i, Y\right).$$

Observe that we also have

$$\mathbf{Ric}(X,Y) = \sum_{i,j=1}^{n} \Omega_i^j(X,V_i)\theta^j(Y).$$

The Ricci transform $\mathbf{Ric}^*$ is defined as the symmetric $(1,1)$ tensor

$$\mathbf{Ric}^*(X) = \sum_{i=1}^{n} \mathbf{R}(X,V_i)V_i.$$

We also have

$$\mathbf{Ric}^*(X) = \sum_{i,j=1}^{n} \Omega_i^j(X,V_i)V_j.$$

The last curvature quantity we wish to mention is the scalar curvature $\mathbf{s}$, it is the function defined by

$$\mathbf{s} = \sum_{i=1}^{n} g\left(\mathbf{Ric}^*(V_i), V_i\right).$$

At this point, this is probably useful to see an example to understand how all this works in action.

Example B.2   Let us consider on the Lie group $\mathbf{SO}(3)$ the left invariant frame generated by

$$V_1 = \begin{pmatrix} 0 & 1 & 0 \\ -1 & 0 & 0 \\ 0 & 0 & 0 \end{pmatrix}, \quad V_2 = \begin{pmatrix} 0 & 0 & 0 \\ 0 & 0 & 1 \\ 0 & -1 & 0 \end{pmatrix}, \quad V_3 = \begin{pmatrix} 0 & 0 & 1 \\ 0 & 0 & 0 \\ -1 & 0 & 0 \end{pmatrix}$$

Let us recall that the following commutation relations hold

$$[V_1, V_2] = V_3, \ [V_2, V_3] = V_1, \ [V_3, V_1] = V_2.$$

For the dual frame $\theta = \left(\theta^1, \theta^2, \theta^3\right)$, we have

$$d\theta^1 = -\theta^2 \wedge \theta^3$$
$$d\theta^2 = -\theta^3 \wedge \theta^1$$
$$d\theta^3 = -\theta^1 \wedge \theta^2.$$

Thus, to find the connection form, we have to solve the system

$$\theta^2 \wedge \theta^3 = \omega_2^1 \wedge \theta^2 + \omega_3^1 \wedge \theta^3$$
$$\theta^3 \wedge \theta^1 = -\omega_2^1 \wedge \theta^1 + \omega_3^2 \wedge \theta^3$$
$$\theta^1 \wedge \theta^2 = -\omega_3^1 \wedge \theta^1 + \omega_3^1 \wedge \theta^2.$$

The previous system admits the unique solution

$$\omega = \frac{1}{2}\begin{pmatrix} 0 & -\theta^3 & \theta^2 \\ \theta^3 & 0 & -\theta^1 \\ -\theta^2 & \theta^1 & 0 \end{pmatrix}.$$

To find the curvature form, we first compute

$$d\omega = \frac{1}{2}\begin{pmatrix} 0 & \theta^1 \wedge \theta^2 & -\theta^3 \wedge \theta^1 \\ -\theta^1 \wedge \theta^2 & 0 & \theta^2 \wedge \theta^3 \\ \theta^3 \wedge \theta^1 & -\theta^2 \wedge \theta^3 & 0 \end{pmatrix},$$

and then,

$$\omega \wedge \omega = \frac{1}{4}\begin{pmatrix} 0 & -\theta^3 & \theta^2 \\ \theta^3 & 0 & -\theta^1 \\ -\theta^2 & \theta^1 & 0 \end{pmatrix} \wedge \begin{pmatrix} 0 & -\theta^3 & \theta^2 \\ \theta^3 & 0 & -\theta^1 \\ -\theta^2 & \theta^1 & 0 \end{pmatrix}$$

$$= \frac{1}{4}\begin{pmatrix} 0 & \theta^2 \wedge \theta^1 & \theta^3 \wedge \theta^1 \\ -\theta^2 \wedge \theta^1 & 0 & \theta^3 \wedge \theta^2 \\ -\theta^3 \wedge \theta^1 & -\theta^3 \wedge \theta^2 & 0 \end{pmatrix}.$$

Therefore,

$$\Omega = d\omega + \omega \wedge \omega = \frac{1}{4}\begin{pmatrix} 0 & \theta^1 \wedge \theta^2 & -\theta^3 \wedge \theta^1 \\ -\theta^1 \wedge \theta^2 & 0 & \theta^2 \wedge \theta^3 \\ \theta^3 \wedge \theta^1 & -\theta^2 \wedge \theta^3 & 0 \end{pmatrix}.$$

It follows immediately that for every smooth vector fields X, Y on $\mathbf{SO}(3)$

$$\mathbf{R}(X, V_i)V_i = \frac{1}{4}X, \ i = 1, 2, 3,$$

$$\mathbf{Ric}(X, Y) = \frac{3}{4}g(X, Y),$$

$$\mathbf{Ric}^*(X) = \frac{3}{4}X,$$

and

$$\mathbf{s} = \frac{9}{4}.$$

Observe that the same computations could have been performed on $\mathbf{SU}(2)$.

There is a natural measure associated with the moving frame $(V_1, ..., V_d)$, the so-called Riemannian measure: This is the measure μ on $\mathbb{R}^n$ given by the density

$$\frac{d\mu}{dL}(x) = \frac{1}{|\det(V_1(x), ..., V_n(x))|}, \quad x \in \mathbb{R}^n,$$

where L denotes the Lebesgue measure.

There is also a natural and fundamental second order elliptic differential operator, the Riemannian Laplacian which is given by

$$\Delta = \sum_{i=1}^{n} V_i^2 - \sum_{i=1}^{n} \nabla_{V_i} V_i.$$

This operator is a Riemannian invariant, that is, it does not depend on the chosen moving frame $(V_1, ..., V_n)$. More precisely, if φ is a smooth map from $\mathbb{R}^n$ onto the set of $n \times n$ orthogonal matrices, then

$$\sum_{i=1}^{n} U_i^2 - \sum_{i=1}^{n} \nabla_{U_i} U_i = \sum_{i=1}^{n} V_i^2 - \sum_{i=1}^{n} \nabla_{V_i} V_i,$$

where $U_i = \varphi(V_i)$. Observe that the sum of squares operator

$$\sum_{i=1}^{n} V_i^2$$

does not enjoy this property and is therefore not a Riemannian invariant. The term $\sum_{i=1}^{n} \nabla_{V_i} V_i$ is actually at the heart of the difference between Riemannian and Euclidean geometry.

One can define Δ in a more intrinsic way. Let us consider the adjoint $\mathbf{d}^*$ of the exterior derivative $\mathbf{d}$, that is $\mathbf{d}^*$ is defined by the property that for every smooth and compactly supported one-form α and any smooth and compactly supported function f, we have

$$\int f\mathbf{d}^*\alpha \, d\mu = \int \alpha \mathbf{d}f \, d\mu.$$

Then, it can be shown that we have

$$\Delta = -\mathbf{d}^* \circ \mathbf{d}.$$

A fundamental property of Δ which stems directly from the previous identity is that it is self-adjoint with respect to μ: For every smooth and compactly supported functions $f, g : \mathbb{R}^n \to \mathbb{R}$,

$$\int f \Delta g \, d\mu = \int g \Delta f \, d\mu.$$

Bibliography

Alexopoulos G.K., Lohoué N. (2004): On the large time behaviour of heat kernels on Lie groups, Duke Math. Journal, To appear.

Azencott R. (1982): Formule de Taylor stochastique et développements asymptotiques d'intégrales de Feynman. In Azema, Yor (Eds.), Séminaire de probabilités **XVI**, LNM **921**, 237-284, Springer.

Bakry D. (1994): L'hypercontractivité et son utilisation en théorie des semigroupes. In Lectures on probability theory, Ecole d'été de probabilités de Saint-Flour 1992, LNM **1581**, Springer.

Baudoin F. (2002): Conditioned stochastic differential equations: Theory, Examples and Applications to finance. Stochastic Processes and their Applications, Vol. **100**, 109-145.

Baudoin F., Teichmann J. (2003): Hypoellipticity in infinite dimensions and an application in interest rate theory, preprint.

Baudoin F. (2003): Stochastic differential equations driven by loops in Carnot groups, preprint.

Baudoin F. (2004a): Équations différentielles stochastiques conduites par des lacets dans les groupes de Carnot, in French, CRAS serie I **338**, 719-722.

Baudoin F. (2004b): The tangent space to a hypoelliptic diffusion and applications, to appear in Séminaire de Probabilités **XXXVIII**.

Baudoin F., Coutin L. (2004): Etude en temps petit du flot des équations conduites par des mouvements browniens fractionnaires, preprint.

Bellaïche A. (1996): The tangent space in sub-Riemannian geometry, In Sub-Riemannian Geometry, edited by A. Bellaïche and J.J. Risler, Birkhäuser.

Ben Arous G. (1989a): Développement asymptotique du noyau de la chaleur hypoelliptique sur la diagonale, Annales de l'institut Fourier, tome **39**, p. 73-99.

Ben Arous G. (1989b): Flots et séries de Taylor stochastiques, Journal of Probability Theory and Related Fields, **81**, pp. 29-77.

Berger M., Gauduchon M., Mazet E. (1971): Le spectre d'une variété Riemannienne, LNM Vol. **194**.

Bismut J.M. (1981): Martingales, the Malliavin calculus and hypoellipticity under general Hörmander's condition. Z. für Wahrscheinlichkeittheorie verw.

Gebiete, **56**, 469-505.

Bismut J.M. (1984a): Large Deviations and the Malliavin calculus, Birkhäuser.

Bismut J.M. (1984b): The Atiyah-Singer Theorems: A Probabilistic Approach, J. of Func. Anal., Part I: **57** (1984), Part II: **57**, 329-348.

Bony J.M. (1969): Principe du maximum, inégalité de Harnack et unicité du problème de Cauchy pour les opérateurs elliptiques dégénérés, Annales de l'institut Fourier, tome **19**, n1, p. 277-304.

Bourbaki N. (1972): Groupes et Algèbres de Lie, Chap. 1-3, Hermann.

Castell F. (1993): Asymptotic expansion of stochastic flows, Prob. Rel. Fields, **96**, 225-239.

Chen K.T. (1957): Integration of paths, Geometric invariants and a Generalized Baker-Hausdorff formula, Annals of Mathematics, **65**, n1.

Chen K.T. (1961): Formal differential equations, Ann. Math. **73**, 110-133.

Chow W.L. (1939): Über system von lineare partiellen differentialgleichungen erster ordnung, Math. Ann., **117**.

Cohen S. (1995): Some Markov properties of Stochastic Differential Equations with jumps, Séminaire de Probabilits, Vol. **XXIX**, Springer.

Coutin L., Qian Z.M. (2002): Stochastic rough path analysis and fractional Brownian motion, Probab. Theory Relat. Fields **122**, 108-140.

Dellacherie C., Meyer P.A.: Probabilités et potentiel, Hermann, Paris, Vol. **I** (1976), Vol. **II** (1980), Vol. **III** (1983), Vol. **IV** (1987).

Derridj M. (1971): Sur une classe d'opérateurs différentiels hypoelliptiques à coefficients analytiques. Séminaire Goulaouic-Schwartz 1970-1971: Equations aux dérivées partielles et analyse fonctionnelle, Exp. 12, Centre de math., Ecole Polytechnique.

Doss H. (1977): Lien entre équations différentielles stochastiques et ordinaires, Ann. Inst. H. Poincaré, Prob. Stat. **13**, 99-125.

Driver B., Thalmaier A. (2001): Heat Equation Derivative Formulas for Vector Bundles, J. Func. Anal., **1**, 42-108.

Dynkin E.: Markov processes Vol.I, Springer, Berlin

Dynkin E. (1947): Evaluation of the coefficients of the Campbell-Hausdorff formula, Dokl. Akad. Nauk. **57**, 323-326.

Eckmann J.P. and Hairer M. (2003): Spectral properties of hypoelliptic operators. Comm. Math. Phys. **235**, **2**, 233-253.

Elworthy K.D. (1982): Stochastic Differential Equations on Manifolds. London Math. Soc. Lecture Notes Series **70**. Cambridge University Press.

Elworthy K.D., Li X.M. (1994): Formulae for the derivatives of heat semigroups, J. Func. Anal. **125**, 252-286.

Emery M. (1989): Stochastic calculus in manifolds, Springer universitext.

Emery M. (2000): Martingales continues dans les variétés différentiables, Lectures on Probability theory and Statistics, Ecole d'Eté de Saint-Flour XXVIII 1998, Lect. Notes in Math. **1738**, 1-84.

Falconer K.J. (1986): The geometry of fractal sets, Cambridge Univ. Press.

Fitzsimmons P., Pitman J.W., Yor M. (1993): Markov bridges, Construction, Palm interpretation and splicing, Seminar on Stochastic Processes, Birkhauser, 101-134.

Fliess M., Normand-Cyrot D. (1982): Algèbres de Lie nilpotentes, formule de Baker-Campbell-Hausdorff et intégrales itérées de K.T. Chen, in Séminaire de Probabilités, LNM **920**, Springer-Verlag.

Folland G.B., Stein E.M. (1982): Hardy spaces on homogeneous groups. Princeton University Press.

Föllmer H. (1981): Calcul d'Itô sans probabilités, Séminaire de probabilités **XV**, LNM **850**, Springer, 143-150.

Friz P., Victoir N. (2003): Approximations of the Brownian rough path with applications to stochastic analysis, preprint.

Gaveau B. (1977): Principe de moindre action, propagation de la chaleur et estimées sous-elliptiques sur certains groupes nilpotents. Acta Math. 139 (1-2), 95-153.

Gershkovich V. Ya., Vershik A.M. (1988): Non holonomic problems and the theory of distributions, Acta Appl. Math., **12**, 181-209.

Gershkovich V. Ya., Vershik A.M. (1994): Nonholonomic Dynamical Systems, Geometry of Distributions and Variational Problems, In *Dynamical Systems VII, Encyclopaedia of Mathematical Sciences*, Vol. **16**, Eds. V.I. Arnold, S.P. Novikov.

Golé C., Karidi R. (1995): A note on Carnot geodesics in nilpotent Lie groups. Jour. Control and Dynam. Systems, 1, **4**, 535-549.

Goodman N. (1976): Nilpotent Lie groups, Springer Lecture Notes in Mathematics, Vol. **562**.

Gromov M. (1996): Carnot-Carathéodory spaces seen from within, In Sub-Riemannian Geometry, edited by A. Bellaïche and J.J. Risler, Birkhäuser.

Gromov M. (1999): Metric structures for Riemannian and non-Riemannian spaces, with appendices by M. Katz, P. Pansu and S. Semmes. Birkhäuser.

Gordina M. (2003): Quasi-invariance for the pinned Brownian motion on a Lie group, Stochastic Process. Appl. **104**, no. 2, 243-257.

Hörmander L. (1967): Hypoelliptic second order differential equations. Acta Math. **119**, 147-171.

Hsu E.P. (2002): Stochastic Analysis on manifolds, AMS, Graduate Texts in Mathematics, Volume **38**.

Hunt G.A. (1958): Markov processes and potentials: **I,II,III**, Illinois J. Math, 1, 44-93, 316-369 (1957); **2**, 151-213.

Ikeda N., Watanabe S. (1989): Stochastic Differential Equations and Diffusion Processes. Second Edition. North-Holland Publ. Co., Kodansha Ltd., Tokyo.

Itô K. (1944): Stochastic integral, Proc. Imp. Acad. Tokyo, **20**, 519-524.

Itô K., McKean H.P. (1996): Diffusions Processes and their Sample Paths, Classics in Mathematics, Springer, reprint of the 1974 Edition.

Jacobson N. (1962): Lie algebras, New York, Interscience.

Kobayashi S., Nomizu K. (1996): Foundations of Differential Geometry, Vol. **1**, Wiley Classics Library.

Kohn J.J. (1973): Pseudo-differential operators and hypoellipticity. Proc. Symp. Pure Math. **23**, 61-69.

Kunita H. (1980): On the representation of solutions of stochastic differential equations. In Azema, Yor (Eds.), Séminaire de probabilités **XIV**, LNM **784**, Springer, 282-304.

Kunita H. (1981): On the decomposition of solutions of stochastic differential equations. In Williams (Ed.), Stochastic integrals, Proceedings, LMS Duhram Symposium 1980, LNM **851**, Springer, 213-255.

Kunita H. (1990): Stochastic Flows and Stochastic Differential Equations, Cambridge studies in advanced mathematics, **24**.

Kusuoka S. (2001): Approximation of Expectation of Diffusion Process and Mathematical Finance, Advanced Studies in Pure Mathematics, Taniguchi Conference on Mathematics Nara, 147-165.

Léandre R. (1987): Intégration dans la fibre associée à une diffusion dégénérée, Journal of Probability Theory and Related Fields, **76**, 341-358.

Léandre R. (1992): Développement asymptotique de la densité d'une diffusion dégénérée, Forum Math. **4, 1**, 45-75.

Ledoux M., Qian Z.M., Zhang T. (2002): Large deviations and support theorem for diffusions via rough paths. Stochastic Processes and their Applications **102**, 265-283.

Lejay A. (2004): An introduction to rough paths, to appear in Séminaire de Probabilités **XXXVII**, Springer.

Lyons T.J. (1998): Differential Equations Driven by Rough Signals.Revista Mathemàtica Iberio Americana, Vol **14**, No **2**, 215 - 310.

Lyons T., Qian Z.M. (2002): System control and rough paths, Oxford mathematical monographs.

Lyons T., Victoir N. (2004): An extension Theorem to Rough Paths, preprint.

Malliavin P. (1974): Géomètrie Différentielle Stochastique, Séminaire de Mathématiques Supérieures, Universit é de Montréal.

Malliavin P. (1978): Stochastic calculus of variations and hypoelliptic operators. In: *Proc. Inter, Symp. on Stoch. Diff. Equations, Kyoto 1976*. Wiley, 195-263.

Malliavin P. (1997): Stochastic Analysis, Grundlehren der mathematischen Wissenschaften, Vol. **313**, Springer.

McKean H. P. (1969): Stochastic Integrals, Academic Press, New York.

Margulis G., Mostow G.D. (1995): The differential of a quasi-conformal mapping of a Carnot-Carathéodory space, Geom. Funct. Anal. **5**, 402-433.

Menikoff A., Sjöstrand J. (1978): On the eigenvalues of a class of hypoelliptic operators, Math. Ann. **235**, 55-85.

Montgomery R. (2002): A Tour of Subriemannian Geometries, Their Geodesics and Applications, Mathematical surveys and Monographs, Vol. **91**, AMS.

Mitchell J. (1985): On Carnot-Carathéodory metrics. J. Differential Geom., **21**, 35-45, (1985).

Nualart D. (1995): The Malliavin calculus and related topics, Springer, Berlin Heidelberg New-York.

Nualart D., Rascanu A. (2002): Differential equations driven by fractional Brownian motion, Collect. Math. **53**, 1, pp. 55-81.

Oleinic O.A., Radkevic E.V. (1973): Second order equations with non negative characteristic form. A.M.S. and Plenum Press.

Pansu P. (1989): Métriques de Carnot-Carathéodory et quasi-isométries des espaces symétriques de rang un, Ann. Math., **129**, 1-60.

Protter P. (2004): Stochastic integration and differential equations, Vol. **21** of Applications of Mathematics, second edition, Springer.

Rayner C.B.: The exponential map for the Lagrange problem on differential manifolds, Phil. Trans. of the Roy. Soc. of London, Ser. A, Math. and Phys., **1127**, v. **262**, 299-344.

Reutenauer C. (1993): Free Lie algebras, London Mathematical Society Monographs, New series **7**.

Revuz D., Yor M. (1999): Continuous Martingales and Brownian Motion, third edition, Springer-Verlag, Berlin.

Rogers L.C.G., Williams D.: Diffusions, Markov processes and Martingales, Vol. 1, second edition, Cambridge university press.

Rotschild L.P., Stein E.M. (1976): Hypoelliptic differential operators and Nilpotent Groups, Acta Mathematica, **137**, 247-320.

Sipiläinen E.M. (1993): A pathwise view of solutions of stochastic differential equations, Phd thesis, University of Edinbhurg.

Strichartz R.S. (1987): The Campbell-Baker-Hausdorff-Dynkin formula and solutions of differential equations, Jour. Func. Anal., **72**, 320-345.

Stroock D.W., Varadhan S.R.S. (1979): Multidimensional Diffusion processes, Springer-Verlag.

Stroock D.W. (1982): Topics in stochastic differential equations, Tata Institute, Bombay.

Stroock D.W., Taniguchi S. (1994): Diffusions as integrals of vector fields, Stratonovitch diffusion without Itô integration, Vol. in the honour of E. Dynkin, Birkhäuser, Prog. Proba. **34**, 333-369.

Stroock D.W. (2000): An introduction to the analysis of paths on Riemannian manifolds, Mathematical surveys and Monographs, Vol. **74**, AMS.

Süssmann H. (1978): On the gap between deterministic and stochastic ordinary differential equations. Ann. Probab. **6**, 19-41.

Takanobu S. (1988): Diagonal short time asymptotics of heat kernels for certain degenerate second order differential operators of Hörmander type. Publ. Res. Inst. Math. Sci **24**, 169-203.

Taylor M.E. (1986): Noncommutative Harmonic Analysis, Mathematical Surveys and Monographs, **22**, American Mathematical Society.

Taylor M.E. (1996a): Partial Differential Equations, Basic Theory, Springer Verlag, Texts in Applied Mathematics **23**.

Taylor M.E. (1996b): Partial Differential Equations, Qualitative Studies of Linear Equations, Springer, Applied Mathematical Sciences **116**.

Thalmaier A. (1997): On the Differentiation of Heat Semigroups and Poisson Integrals, Stochastics and Stochastics Reports, **61**, 297-321.

Watanabe S. (1984): Lectures on stochastic differential equations and Malliavin calculus, Tata institute of fundamental research, Springer Verlag.

Yamato Y. (1979): Stochastic differential equations and nilpotent Lie algebras. Z. Wahrscheinlichkeitstheorie. Verw. Geb., **47**, 213-229.

Yosida K. (1952): Brownian motion in homogeneous Riemannian space, Pacific

J. Math., **2**, 263-296.

Young L.C. (1936): An inequality of the Hölder type connected with Stieltjes integration. *Acta Math.,* **67**, 251-282.

Zähle M. (1998): Integration with respect to fractal functions and stochastic calculus I. *Prob. Theory Rel. Fields,* **111**, 333-374.

Index